WHAT IS TIME?
WHAT IS THE ORIGIN OF TIME
and
THE SENSE OF DURATION?

WORKS BY SAMUEL K. K. BLANKSON

The metaphysical Foundation for Physics;

Why time is not a natural phenomenon;

The Mathematical theory of time;

How old is the universe by our time;

The Einstein theory of space-time without mathematics;

Dead End (a novel);

Time in Science and Life---The greatest legacy of Albert Einstein;

How religious scientists play down the greatest of Einstein's achievements;

The coming revolution in physics;

Time and the Application of time;

The Logic of time in the universe;

Philosophical Essays;

On the Nature and Passage of time and 4-D Geometry

Past, present and future as time in the age of science

WHAT IS TIME?
WHAT IS THE ORIGIN OF TIME
and
THE SENSE OF DURATION?

Samuel K. K. Blankson

PRACTICAL BOOKS
2017

Copyright © 2017 by Samuel K. K. Blankson

All rights reserved. This book or any portion thereof may not be reproduced or used in any manner whatsoever without the express written permission of the publisher except for the use of brief quotations in a book review or scholarly journal.

First Printing: 2017

Paperback: ISBN 978-1-326-99208-8
Hardback: ISBN 978-1-326-99295-8
Ebook: ISBN 978-1-326-99211-8

Published in Great Britain in 2017 by
PRACTICAL BOOKS
Blankson Enterprises Limited
92B Howson Road, London
SE4 2AU

www.practicalbooks.org

Dedication

To the memory of H.A. Lorentz, Albert Einstein, Bertrand Russell and A.N. Whitehead among the brilliant thinkers who changed the world with the notion of secular time.

Contents

FOREWORD XI
PREFACE XV
    ESSENTIAL DOCTRINES xli
INTRODUCTION 1
CHAPTER ONE: THE SCIENTIFIC NOTION OF TIME 61
    ASTRONOMY AND TIME 81
    WHY DURATION RESULTING FROM CONTACTS IS THE CAUSE OF THE SENSE OF TIME 84
    WHY SECULAR TIME IS NECESSARILY DISCRETE 91
    WHAT A MOMENT MEANS 101
CHAPTER TWO: THE NATURE AND ORIGINS OF A TIME SYSTEM 121
CHAPTER THREE: THE MINKOWSKI EQUATION OF SPACE TO TIME 127
CHAPTER FOUR: THE FOUR AGENTS THAT CAUSE TIME INTERVALS 151
    The Order of Time Seen As a Matter of Arithmetic 164
CHAPTER FIVE: CONDITIONING THE HUMAN MIND FOR TIME 183
CHAPTER SIX: THE NATURE OF TIME BEFORE AND AFTER EINSTEIN 195
    (a) - TIME BEFORE EINSTEIN 196

- (b) - WHAT IS SECULAR TIME?...207
- (c) - TIME AFTER EINSTEIN 210

CHAPTER SEVEN: ENTROPY, GRAVITY AND TIME......217
CHAPTER EIGHT: ENTROPY AND TIME.........................241
CHAPTER NINE: GRAVITY AND TIME.............................248
CHAPTER TEN: PHILOSOPHER/SCIENTISTS.................257
CHAPTER ELEVEN: THE STATUS OF EARTH TIME IN THE UNIVERSE...281
APPENDIX I: TIME AND QUANTIFIED TIME OR THE PASSAGE OF TIME...300
WHAT IS MEASURED BY THE CLOCK? 309
APPENDIX II: THE PRINCIPLE OF MATHEMATICAL EQUIVALENCE 314
APPENDIX III: WHY SPACE ON ITS OWN IS NOT "SPACE-TIME" 323
APPENDIX IV: THE MISCONCEPTIONS OF TIME IN RELATIVITY...330
CONCLUSION 343
REFERENCES 368
INDEX...372

Foreword

Leaving aside gravity, matter and light as far beyond human comprehension, there are only four things in the whole universe that require philosophical interpretations of ultimate reality, from the point of view of human welfare. They are: Space, Time, Consciousness and Death. Philosophy is important because we need to know the ultimate causes of reality; they agitate our minds and also they form the basis of all material subjects for scientific study. Space is obvious because it's what we live in. Time is the duration of all aspects of life, and it became human in origin when cosmic time was abolished due to the discovery of local time. Consciousness is the human awareness of the existence of all matter, but Death has only got to be proved as to whether it is the total end of any human being. The billions and billions and billions of other objects found in, or created by, planetary conditions (including the 'Day&Night' systems) are not metaphysically known in the universe---being temporary, planetary events many of which are man-made---and therefore do not count in a philosophical discourse of primary causes. We can't use them to explain what is unknown to them. This is one of the principal laws of nature frequently breached in religious discourse, probably it applies to the debate about gravity and the quantum as well, since the quantum is the basic matter in our observable universe, whereas gravity (whatever the mathematics) is only the effect of bundles of matter, a completely different category of existence... It is futile using something to explain something unknown, either way, to that something.

This book is dealing only with time. However, time, as the duration of existence, has been mathematically subdivided into manageable units we carry with us wherever we go as 'time units by the clock', from the year down to the second, the latter being our SI of time. Without the clock we would not know when to do anything, especially in chemistry and physics. This makes the philosophy of mathematics our most important subject, provided it is completely stripped of fantasy, mysticism and eccentricity. Time units are shorter units of duration for the regulation of all activities---hence the complexity of what is known as secular time without mythologies, religion or any fallacious arguments. It may even be wise to call it the scientific theory of time, because it was Albert Einstein who initiated the debate to abolish cosmic time with the enthusiastic approval of Bertrand Russell and Professor A.N. Whitehead, to whose memory this book is dedicated. By logic, mathematics and phenomenal brain power, these three men deduced accurately that, "There is no longer a universal time...", and that what we call time is man's peculiar way of looking at the world. The universe at large is not governed by time but by random chance and gravity, which is usually overlooked---therefore the most mysterious thing in the cosmos is the human brain which invented time. "The imagination", according to the American poet Wallace Stevens in Adagia, "is man's power over nature", yet nobody can ever find out where the brain got its powers from by using the same brain; for another of the unbreakable rules in epistemology is that nothing can discover or make itself. It can only make something else: using something to discover the nature of that same something is not an acceptable method in the theory of knowledge---hence the infinite variety of creation, and probably the reason why we exist at all, as matter creates different compounds on and on and on. Even identical twins are different, namely, as two bodies out of one egg.

Preface

The days are not part of time in astronomy. There is only one day all over, and here is the reason: as human beings we are of course complete strangers on this planet with absolutely no idea where we came from and why we are here only to die eventually, sometimes painfully, and disappear from the earth. We notice, however, that among the billions and billions of objects in the world two of them appear to be so close as to be practically inseparable. They are life and time; so we are inclined to regard them as the original metaphysical creations on earth, all other things and events being generated by or on the planet; being locally generated, we know how they came into existence; some of them (in agriculture) are even cultivated by us. So how can we use them in attempts to explain reality? The irony is that the day and night system by which we organize our lives is one of those things generated by the planet's rotations and not known outside the earth; thus it is irrelevant in the discussions of time and ultimate reality, since outside the earth there are no days just as there are no trees on the sun.[1] We and our planet plus the billions of objects and events on earth even taken together as one huge lump, is too tiny to count in the universe anyway. One implication is that worshipping any object or 'Being' on this planet for cosmic favors is a waste of time, energy and resources; we are too small to be noticed anywhere. The gravity of the planet may be felt in the journey round the sun, but the rotations (for day and night) as a source of influence can only be something of a joke in the universe---I call it the planetary dance of no metaphysical relevance!

So there is only one constant 'daylight' everywhere in the whole universe. I repeat, there are no days and nights in any part of the cosmos. The stars do not go to sleep. With reference to time, in the primitive days before the rise of science we simply noted the intermittent darkness accidentally created (not planned) by shadows of the earth, turned it into a religion, and invented an artificial language---since all language is artificial---so as to call it 'Days marching on as an indication of the march of time to Judgment Day'. This is completely illogical, but it has (unfortunately) misled mankind into believing strongly that time moves across space to a providential predestined, preordained and predetermined end without fail, and on the way there kings like Henry VIII with all their faults must be worshipped till their glorious deaths and burial with

---

[1] There are objects on the sun we don't know about; and there are things in the world not known outside. They are called 'matters of internal chemistry'. We can't use them to interpret reality---that is why the nights do not count in the interpretation of time.

servants to serve them in the afterlife.[2] This has distorted and caused damage to human thought for centuries and is so strongly and religiously held that it may prove impossible

---

[2] In fact, time moves together with its allotted space or a portion of space converted (with mathematics) to intervals as cultural units of reality, but its advance is not physical that is why things remain the same for years and years, especially buildings even though time has moved on for centuries—i.e. church buildings, especially in Oxford! The reason things remain the same is that time is only a concept---which is where the mystery of time begins. It is a concept mathematically linked to the orbits of the earth round the sun; as such, time cannot ride over space or be the same as space, but moves with it in the mind only, exactly like arithmetical calculations: thus we can say, "James was here ten hours ago." The time has moved on in numbers but we are still in the same place, because the time is conceptual. That is why a precise number of seconds (**in temporal terms this means a precise number of waiting periods called 'seconds'**) equal to one complete orbit of the sun---and we start another year. Thus a mathematician can calculate the time even if blind. The Minkowski theory was a great effort, but not really true. Space-time means time is based on space and could not exist without it, not that they constitute one entity. Time without space ends in divinity; it's only with space it becomes truly scientific or objective. Mathematically all units of time are multiples of the second. The SI of time was chosen for both mathematical and metaphysical reasons, even psychology is included. Time is more complex than we think. But the days, weeks and so forth are customary practices based on religious beliefs. This is the most scientific theory of time. However people don't want to hear of it, as they prefer the Day of Judgment system of time of which somebody has written A Brief History, made into an all-time Bestseller by the religious people. On the contrary, scientific time has no history, one moment and is gone to be succeeded by another moment, as Professor Whitehead called it, 'a sequence of non-interacting moments'. Only events have history not time; the time is grafted to the events. Time does not move across space; it replicates, like the years. If the religions do not get in the way, science could be more successful than it is now. Google and others know this so they're trying to advance little-by-little though functional devices.

to dislodge from the mind. Yet, in fact, no star can produce repetitive night-times, least of all the sun by which we live. Having lived under the sun for so many thousands of years, we know it cannot perform such a miracle. There is no doubt that time in people's mind is all related to the false idea that time is marching through nature, particularly because of what they see as successive 'days'. Yet there are no days either in astronomy or metaphysics. I am not surprised that Professor Eddington used strong language about this, for it just is not part of ultimate reality.[3]

The noble founder of astrophysics, one of the greatest scientists of all time, knew more than anybody else that the regular day-and-night system is a human concept base on a fallacy (another name for religion), and does not count in the reckoning of time, since it is caused by the mere passing shadow of the planet. There is nothing in metaphysics that we can only do by night and not by day, yet only that would make the night time essential aspect of reality. Furthermore, this book is not discussing philosophy in the traditional sense, concerned with arcane subjects like God, German Idealism, Plato's Idealism or Moral Theology, the

---

[3] Reality consists of layers, as the study of physics shows---from the light we see to its accumulated bulks, such as sub-atomic matter, ordinary material objects, planets, moons, stars and beyond. Planetary days and nights (including billions of people being born, living and dying,, trees, animals, atmosphere, rain, clouds, the grains of sands, rocks, shrubs, etc, and their numerous interactions) are on our cultural (or different) levels of reality; they are not part of ultimate or metaphysical reality. Time, life and death occupy the apex of metaphysical reality and all human concerns---very mysterious, unavoidable and ultimate nature unknown or unknowable..

Cartesian Dualism and thousands of vague topics. I can't even understand why intelligent scholars bother with such subjects of pure fantasy, the main reason philosophy is not taken seriously by the public. Rather the book is an ordinary man's discourse that recognizes reality as defined in physics only, particularly Quantum Physics and the standard model, for the simple reason that scientists know enough of physical reality to be capable of destroying the world. Indeed, to my mind, although the quantum theory is strange in many ways, yet it made the logical analysis of physical reality consistently structured (as accumulations of quanta) and therefore easy to tamper with, and there are no lack of monsters all over the world who may want to do so for a variety of reasons. Let me clarify that, although time in the clock is man-made, the units of time are metaphysical, they are the culturally manageable units of reality, cleverly mechanized in the clock, that we apply to events as 'time'.

The whole of reality is time; that's how closely life is associated with time; every moment of existence is a moment of time. Our consciousness of existence is the time of our existence. And this is after all religious notions of time have been discarded. Units of this reality are what the clock provides through mathematics; but of course if there is no longer a universal time then time is man-made. And we know that a second is a portion of physical reality because it is derived from (actually, physically, part of) the space traversed by the earth round the sun. If we place points continuously in seconds each, by the time we get to 31, 536, 000 one year will be up, for a second is equal to an amount of space directly derived from the space round

the sun. This is the best logical definition of time upon which the whole book is based. The shallow academics who reject this are deluding themselves---especially those at CUP who claim to be the cleverest on the planet and yet worship Wittgenstein as a great philosopher.

The day and night system is not in the same class as a second, because it is not a portion of physical reality but just a shadow (like a cloud) over reality which is soon dispersed to leave reality intact. The nights change nothing. Of course, we organize out lives round them, but we do the same with trees, rivers and seas etc. The confusion arises from conflating time, as a period of waiting, with its use, its origins and its interpretations. One of the purposes of this book is to suggest how we might (or should) separate them intellectually even if not culturally---alas, nobody can divest mankind of its religious believes which are all based on the nature, provenance and future of time, regardless of what the logicians are saying in their mighty tomes.

■■ı

Thus I begin the book with the definition of time as a whole to avoid confusion, because what is definitely contentious in the whole debate about time (and impossible to prove), is the traditional/religious notion that the passage of time is the passage of existence, or the passage of existence is synonymous with the passage of time. The problem is that there is no other explanation for the passage of time, and simply because we age 'over time', shallow thinkers in science, philosophy and religion have concluded that existence **(meaning 'contact with**

nature as experienced'⁴), moves irreversibly forward with time, 'as we age', hence the 'Arrow of Time' theory. The idea is strictly derived from the irreversible ageing process; and man has been ruled for thousands of years with this myth so fanatically that I get no one willing even to look at the contrary ideas I am putting forward. Hence I think what we're talking about needs to be defined first. But, in any case, does the old idea make sense? Do the houses and everything move in tandem with time, so that each time the clock ticks the seconds successively existence moves with them? Is the motions of existence (or 'Beings') in their multitudes controlled by time? Again, is time logically a concept formed by man to apply to the world and make it habitable, or a separate physical entity running all through the cosmos and the same everywhere?

---

[4] This means man against nature. We exist and we are different. Descartes' formulation was wrong. The sense of being as against the sensing of other beings is what matters as the beginning of all knowledge, individuation, society, ethics, et al. This is what metaphysics means or should mean, and can be developed to become "The science of life, religion and Democracy", which Descartes never considered, namely,' you are not alone and need the others'. To exist is to be aware of the world as also in existence; it is not different from experience; but it can be seen as 'contact' with the external world, which would include even contacts in the womb---see below. Thus if experience is existence and the passage of time is the passage of existence, then human contact of any kind is time, and it cannot move; what is moving are the repetitive cycles we use to reckon time. To say these cycles constitute the motion of time *itself* is a delusion; with the aid of mathematical thinking the cycles can only show us how much of time is passing, has passed, or will 'eventually' pass by. The true nature of time remains unknown---there are some suggestions about that in this book, though.

Even in physics and biology not all molecules move in unison; and other motions all over are wildly different. On the contrary, I argue that the passage of existence is chemical or occurs through accidents and that time, seen as synonymous with existence or contact with nature, does not move,[5] it is merely conceptual; it advances through the procession of its constituent units----and that all time consists of units because the units are fractions of the yearly cycle, which is determinate so that even for the year itself to advance it has got to replicate not that it flies through the air. This means time does not move but replicates its units to pass by or advance in procession, and not in the form of a thread, either, but unit by unit---like the years, the seconds, the hours, et al. This is the philosophy that the book is planned to explain down to the

---

[5] Ageing with time is in the mind only, because ageing is chemical and has nothing to do with the motions we use to track time, for time, after all, is unknown and unknowable; we can only trace how it passes by. What we know as the movements of time are rather the motions we use to reckon how the time is passing by---we count the years as the number of time units (years) that have passed by, yet the years are physical not temporal, and counting the years is really counting the rates by which the time is passing by and never what it is, and what applies to the year applies equally to all other units of time mathematically conceived as fractions of the year. Of course it is perfectly reasonable to assume that the earth's orbits are time moving forward with the earth's entire contents, including ageing. The problems is that the earth's orbits are physical not temporal and nobody knows how much time they displace, only that time is passing---see below. Everything we use to reckon time can only show that time is passing and never what it is, and mistakenly we call the motions of the cycles we use as 'the movements of time', but at least we don't need to find a theory to account for the passage of time. There is no passage of time; what we call time is how it is passing by.

level of ordinary readers' understanding: traditionally time is known as existence so that the passage of time is the passage of this existence. I also think all time is known in units as they are fractions of the determinate yearly cycle. Given these two propositions, what we have to decide is whether time, as existence (or like existence), is passing by physically. But first, if time is existence then it means time derives from human contacts with nature (because 'contact' is another name for experience or how we perceive existence), but these contacts cannot be thought of as being in perpetual motion like the traditional notion of time. Otherwise how could we get the steady images of reality to live by? It's obviously logical to assume that experience means contact with nature, and that the experience itself does not move; other objects can be experienced as 'moving' or as stationary, etc., within that one experience but that is a different aspect of time and perception. On the other hand, if you touch an object, the tactile sensation itself does not move; but you can count other cycles, and say the touch lasted so many cycles---- **that is time and also does not move. The repetitive cycles like the years we use for reckoning time or the duration of the touch do move, but not the time as a portion of the repetitive cycle. That portion as a unit of time, once established, does not move. It is the motions of the cycles we call 'time intervals' (or intervals between points). The intervals themselves do not move as they constitute the time or sense of duration.**[6] However, lacking any logical

---

[6] Ten intervals means ten minutes---that does not move. The ten minutes is established and cannot change. We apply it as a given duration, or a

explanation for time, we've mistakenly assumed that the movement is that of time itself; no. It is of the cycles used to establish the time in units. Time is merely a conceptual accessory that can only advance through the replication of its conceptual units in perpetual procession, ultimately with the speed of light (because of vision), since all time is known and used only in the units we assign to divide reality for our convenience.[7] Culturally time is useless until it is quantified, as explained below in Appendix 1.The fact is we can only count time in units, not in any other way since silent time is not usable.[8] And, obviously, the units can only advance by themselves---i.e. unit by unit. Thus in the mind all time is known and used only in units or Whitehead's 'moments': the gaps between points. The whole of human existence is based on the use of time, and it turns out to be

---

concept, which we know as 'time, but, as time, it means time does not move. Like a badge; we apply it as a whole and cannot move. What created it elsewhere could change, but not the time otherwise we couldn't get steady concepts to live by.

[7] An equation incorporating MC^2 is hidden in this sentence, waiting to be discovered by some clever mathematicians! That these ideas about time have led to all human thinking, science and civilization can be used to counter all religion, mysticism and the Pythagorean notion that human knowledge is a travelling Circus going round and round the cosmos. It is also interesting that deep thought about time should yield such momentous benefits.

[8] This is one of the new ideas in philosophy that needs to be stressed again and again to overcome ingrained instincts about human values---namely, the death of silent time. Silent time is simply being there plus motion without the intellectual use of points, the crucial part of being human. We used to be told that it is 'points and instants' but that is wrong. It is the intellectual use of points and motion as the instants arise from moving from point to point.

merely the gaps between points, but it means we're getting there, getting closer to the nature of life! That is what the study of time should begin with to avoid most of the mystery with which it is encapsulated. Otherwise with any experience how could we know its duration in units, say, for five or ten seconds? Yet the point is that the seconds come from the yearly cycles. The yearly cycles give us all our units of time to apply to events so that our time is not arbitrary but based strictly physically on the orbits of the sun.

So there are three aspects of time: first the contact with nature (or experience), otherwise there is no need for time. Next, we create repetitive cycles with 'the intellectual use of points' and divide them into periodic units---time to you and I---as the experience, event, action, goes on. This is what misleads people to make them say 'as time goes by' or 'goes on'. In fact, it is 'as life goes on' or 'as the repetitive cycles used for reckoning time go on and on and on indefinitely' so long as the sun is there. The units of time obtained from the orbits of the sun are applied successively to events (or whatever) as they go on. If we have a 4-hour speech at the UN (Castro style), the cyclical hourly units would move (or be applied) four times to the actual image of that one continuous speech. This is an example of the cycles we use to reckon time moving rather than the time or the event itself being in motion. It is the cyclical motion we call time, and it is indeed **the time (all we can ever know of time)** and it moves: it is the hour and it moves four times, not the event. What real time is we do not know, and it does not appear to be knowable. We use cyclical units to show how much of it is passing by, being utilized or expended. Since all this happens simultaneously and very rapidly we are confused and call it time for short.[9] What applies to the hour also applies to every unit of time. That is how we invent time units out of the space traversed round the sun; and because the cycles are incessantly repetitive, duration is forcibly imposed, making time preciously and oppressively limited, because the cycles are, of course, determinate; yet they create all the environmental conditions for sustaining life on the planet as it is influenced by the sun. So the short message is that to live on this planet we have got to obey its time---as always, Einstein was right: every planet has got to have its own suitable time system. That is the philosophical, scientific and logical origin of the sense of duration and the 'universally acknowledged' oppression of time. 'Time indeed waits for nobody'.

It's all logically structured perfectly by the human mind. To my mind this is no more mysterious than the creation of language; yet no one has seriously suggested that language is divine. Duration is basic and it is the product of points; time is used to divide duration into convenient units. The year, for example, is one long period (or duration) divided into 31,536,000 sub-units called seconds. Everything is based on our orbits of the sun.

---

[9] Not only confused but completely lost through ignorance: before electricity, the quantum and QED, we did not even know how images are formed and relied on the Platonic fiction or religious myths, thus making writing about time the most painful thing in the universe. It's taken me more than fifty years to achieve this little, and yet still nobody will even condescend to look at what I have done from a distance.

The third aspect of time has already been covered in the above explanation, namely, we count the cyclical units with their imposed durations as units of time for application to reality. To take time as 'just is' without this analysis will, of course, make it seem pretty mysterious, and the religions made things worse by building all theology on the conundrum of time.

And let me also stress the greatest intellectual problem again: time does not advance by moving physically but rather by replicating its units numerically----the years increase in numbers to pass by. The reasoning is not that complicated, except that traditional concepts of time have dulled our senses. Our units of time are fractions of the repetitive cycle we call one year. As fractions of an incessant cycle they are fixed; they cannot be changed by us. They are, in a sense, 'given' as supposed by the religions. Also we think that time is marching when the clock is ticking; but in fact the clock only repeats the predetermined units of time. The time is 'given in fixed units'; we merely apply the given units to events.[10] The clock does not march with time; it merely repeats the units of time programmed into it: it does not march from one to two and three and so forth. It repeats the seconds, second,

---

[10] I think we should regard time as given in the religious sense except that it is our brain and not God who made it for us. That may be the best way to understand secular time as given or gifted by our orbits of the sun. The obvious advantage is the satisfaction of knowing that the nature of time can be logically deduced from known facts, physical conditions and philosophic intuition rather than blind faith in baseless theology and Wittgenstein.

second, and second ad infinitum. Time advances through the replication of its units. It does not march as the units are metaphysically fixed and cannot move on in application so that reality would move with the ticking of the clock. We stay where we are. Reality does not move with time; it moves by its own chemistry or through accidents such as are caused by storms. The units of time (as 'a sequence of non-interacting moments') move digitally in constant procession up to the speed of light due to the manner in which they are created as fractions of strictly determinate repetitive cycles. The reason we have time increasing even as we are applying it (which gives the duration of events) is that the units of time are forcibly obliged to multiply constantly. We apply suitable digital units (always in motion or multiplying unit by unit), to events and wrongly assume that the events are moving with time, or that the time is causing them to move on[11]---yet the events and we remain where we are, and grow or move mostly chemically whilst the time units continue to multiply to pass by. In theory, time is a concept we apply to events. It advances only

---

[11] It is only in application we notice time; the reason we assume that it is part of reality and even moves inseparably with it, because time, events and existence are ceaseless and occur simultaneously (hence the notion of time as the passage of existence; one by one we can dismantle the structure of traditional time.) Without application time is mere infinitely variant motions to and fro in the wild, not in the mind. In the mind it is always in application---thus the misconception that time is the prime mover and moves with all existence. All regular motions can be used to reckon time, for the nature of time is unknown, we count cycles as the rates of passing time, which is what we call 'time'; and we choose the earth's motions because of unavoidable astronomical influences on the planet.

through replication or arithmetic; and while it has to be applied to all events, it has no mechanism to move or replicate in tandem with actual reality. That, in any case, is the philosophy. While nobody knows the true nature of time for sure, logic, as the ultimate arbiter of truth, nevertheless dictates the above as one man's fallible suggestions. I define time as a period of waiting, work, motion and space traversed, delay and stoppages, growth, decay, ebb and flow and any activity whatsoever. Or even silence, stillness, standing, sitting and lying down; consciousness and unconsciousness are also covered by time.

But now comes the literary mechanics: I can confirm that writing about time is akin to discussing religion; opponents are hostile not charitable. So I will begin with a few preliminary lines about the difficulties of discussing a critical aspect (indeed, the basis) of life, even though I am criticized on the internet that my country is not known to be capable of producing any philosophers. I do so also in answer to several other queries about failing to submit my work for peer reviews and the specialist journals. My excuse is that this is not science to make peer review obligatory. There are those backroom mathematicians Newton said he's afraid of, armed with massive theories about the multiple dimensions of space, past, present and future, human destiny in an endless universe and all the rest of it, and I will not allow them to put me down---even without knowing what they're really talking about. My sin amounts to merely asserting that at least the problem of the passage of time has been solved, if no more, but not from any one thinker's ideas. We have to go far back to relativity, Russell and Whitehead. An additional sin of mine is that I do not subscribe to the multiple dimensions of space in mathematical physics. Because of time, I believe only in the visible 3+1 formula---so did Einstein upon whose supposition my theory is based.

But of course I have to plead for understanding. Time alone of all subjects, due to the mystique and close association to life, is almost impossible to discuss without meaningless metaphysics, religion, revelation and questions of human destiny.[12] As a result, it has no background literature, post-relativity time much less, yet that is my specialty. Generally speaking, people and thinkers are either afraid of time or incapable of discussing it rationally.[13] Instead we get isolated, lopsided or irrational comments that fail to answer the most important questions.[14] There

---

[12] I am not qualified to discuss human destiny. I am an ordinary person, born, raised and destined to die and disappear altogether. That is all I know about human beings and nobody has found anything else, it's always dreams, dreams and more dreams called 'revelations' to gain advantage in life.

[13] I have come across people who claim to applaud my work but will not sign me on as my agents. Another agent begged me to go elsewhere because he did not understand mathematics! Even the detractors attest that my work is readable.

[14] Martin Heidegger's writing is a notorious example. Only Leibniz and Bergson, (plus Russell and Whitehead with the benefits of relativity), have ever conceived reasonable ideas about time for the rest of us to build on, commendably supported by the genius of Professor Eddington, as mentioned in the book. To reject these ideas is a disservice; pleading ignorance of mathematics is no excuse as they are presented in plain language even in The Mathematical Theory of Relativity!

are, for instance, no experts beyond the chronometer or the retort that time just happens to be there, which I reject as unhelpful. In all history only Einstein's views about time are provable (even more recently the GPS use of time has also added another evidence that time is an additional coordinate vindicating the 3+1 formula.) Of course this makes the Einstein theory the greatest revolution in human thought, for time is the biggest conundrum of all, second in importance only to life, and yet mystifyingly time seems inseparable from life or sentience.

Nevertheless, philosophically, all Einstein did amounts to merely showing that time is neither absolute, fixed nor generally covering the cosmos such that a second here is a second everywhere else; or that, there are as many times as there are inertial bodies. He could not tell us what it is, and part of my theory is that no one can ever reveal the true nature of time. Intellectually all human beings are fallible. I am just asking these mundane questions: even if time 'is just there', what is it that is just there, how did it begin, and how does it pass by; also what is the meaning of past, present and future plus a bagful of other questions? And the fact that I originally came from Ghana which is held against me is mere red herring---old enough to brush it aside.

I am asking mundane questions about time that everybody can understand. The Cambridge University Press tell me, insultingly, that my work is below their highbrow standards, as I seem to be writing popular philosophy, whatever they mean by that. But to me this is normal; it is another insult to a black man. I get them every day in the white man's country; so I meekly reply that I write in a manner likely to interest ordinary readers because we all use time. It's different with other theoretical subjects like Quantum Physics, I agree. But to seek to write about time only for the dons is unfair to the rest of mankind who do not teach in a university yet use time and want to know what it is, since the academics have failed over many centuries to reveal its true nature. They couldn't do so even when they're serious and dedicated; now they don't even try because they're too busy making money from the media.

Now logic, obviously, is the ultimate arbiter of truth, and the most logical (or evidence-based) definition of time after relativity is 'A PERIOD OF WAITING IN THE MIND ABOUT ANYTHING WHATSOEVER EVEN IN DREAMS', a conceptual not a natural phenomenon, but based on known natural and physical parameters so that the notion of time could not be conceived without them.[15] This alone is a mouthful, yet that is only in logic or rational thought. There is chronometer, the Minkowski fiction, post-relativity-time, mythology, organized religion and Day of Judgment, philosophy, tradition, the clock, astronomy, cosmology, black magic and black holes (I put them together), voodoo, linguistics and the practicalities of time yet to consider in any attempt to define time. Nothing in the whole universe is more complicated than time except life itself; we can't even live our lives without it. The fact

---

[15] Einstein proved by experiments that time is neither absolute, fixed nor generally covering the whole universe so that a second here is a second everywhere else---see below. The GPS's use of time is one confirmation of this.

that it occurs even in dreams is proof that it is basically conceptual---a concept formed of all objects and all events perceived and as they react to one another, the reason perception or contact appears to be the basis of the sense of time, beginning even from the womb. This definition alone (as detailed in this Preface) is longer than some of the chapters in the book; but we have to do that, for when we discuss time, we are indirectly discussing life as well, the most momentous affair in human existence overall that requires detailed and cautious, even tremulous or nervous approach: to live is to expend time, but what is it and where does it come from?

For the individual writer struggling to make sense of all this after relativity, logic is the only guide, without the recklessness and vanity of claiming to know anything for sure. The closest we can get to knowing time seems to be how it is passing only---the years, for instance. They are passing one by one successively; if that is time passing by then that's all we can ever know about time: how it passes by not what it is. But to regard the passing days as evidence of the passage of time is a mistake because in astronomy there is only one 'constant' day or daylight, as explained in the text below: the sun never goes to sleep.

Even then, without incorporating 'ds^2...' and the '$s=ct$...' equations in the definition, mathematicians will argue that the mathematics of abstract and metric space has been ignored and therefore the reasoning is flawed. I know they'll say that. These two (Minkowski-inspired descriptions of the universe) are well-known areas of science where mathematics becomes really magical. However they've not been overlooked; but in the study of time, the same time that jungle people use in their kitchens without calculators, details of the Minkowski universe do not matter; everything is incorporated in the creation of the clock, and they do have clocks even in the jungles. The clock too is accounted for by the answer to the Russellian question: what is measured by the clock? The answer, of course, is the metric space round the sun traversed by the earth, and measured in units as relations, intervals, or contacts between points. This is the ultimate of abstraction in the study of time---as far as mathematics can go in the definition of what gives us time: points, the intellectual use of points, and the intervals created by moving from one point to another (exactly the way we get the year as a unit of time), is how time must be defined in mathematical physics. Contacts or perceptions even in the womb are also 'relations between points'. Thus time is not all mathematical and it's not wholly physical either; human elements are required because somebody must be there to set the points for their calculated relations to become units of time in the human mind---or there will be no time. This is the most logical definition of time I can think of. It is complex because time is very odd indeed.

Strangely, the universal human feeling that time is cosmic (whatever 'cosmic' means) dies hard. It derives from time's mystifying nature. There are two theories that I am aware of: (1)"That it is part of life and originated with it together and nobody knows or can ever know from where they came". Yet the religious interpretations are no longer credible due to the rise of science. (2)"That man should not probe time too deeply as no one knows much about it because it was created by whoever created the universe and life, and we little minions should just be glad to have it for use and not worry about how it came to be there".¹⁶

However there are problems even just accepting it as it is. For a start, we need to define it logically so as to be able to control it. In all human affairs the inability to control an aspect of nature is what we dread the most. The traditional/religious view is that time consists of fixed periods, general and absolute, so that a second here is a second everywhere else. But then Einstein and Lorentz discovered that it is not so: neither fixed nor absolute and the same everywhere but rather varies from place to place and also anybody can create his or her own time---for that is how we get the years: as periods between points. The year is gained as an interval from one point to another

---

¹⁶ Let me explain the cryptic point above that I put black magic and black holes together. The reason is that despite the high-brow guff about meaning and purpose for life and the universe at large, and where we go when we die and all that sort of thing, in logic it is obvious to me that the universe exists in a swirling mix of comings and goings without plan, purpose or meaning, precisely like all the mumbo-jumbo in Black Magic. Life is nowhere guaranteed. Without good care we die soon after birth. The greatest mystery is our intelligence; and what I suspect is that some crafty writers are misusing it to choose black holes for speculation, knowing that verification is quite impossible---and still have the gall to demand the Nobel Prize because they've made a discovery about black holes. Who cares---or what for? Even then how does anybody verify any proposition in a black hole? Since the days of the great and noble Bertrand Russell, the whole World's intellectual output has declined, chiefly in quality, as writers have grown pretty greedy due to the media frenzy and cheap money from the electronic revolution, especially mobile phones. Now they demand the Nobel as well, and get it through diplomacy. I think we're all going downhill, intellectually. Even science is affected.

point. The important thing is that the year is determinate, meaning it is one period repeated over and over again with points as the passage of time. This is the most important logical point in the study of time. It means mathematics comes in, for after all we know and use time only in units, suggesting that time is human and discrete (as discussed below), giving rise to a whole lot of rational, non-religious theories about its origin, passage and destiny---if there is anything like destiny at all about time.

I cannot claim to know more about time than anybody else, except to point out that the scientific explanation sounds convincing and makes it possible to know how it passes by---since the years give us a clue—i.e. the years increase in numbers to pass by. In this sense the passage of time becomes the ability to count its units on and on. So is it possible that all time (in units) has no arrows to use to pass by but merely increase in numbers just as the year increases in numbers to become centuries?

So then, we just have to begin this daunting task with the views of Professor Sir Arthur Eddington because he sets the Einstein theory of time in an admirable perspective. "Prior to Einstein's researches no doubt was entertained that there existed a 'true even-flowing time' which was unique and universal...Those who still insist on the existence of a unique 'true time' generally rely on the possibility that the resources of experiment are not yet exhausted and that someday a discriminating test may be found. But the off-chance that a future generation may discover a significance in our utterances is scarcely an excuse for making meaningless noises..."[17] This was stated by Professor Sir Arthur Stanley Eddington in his monumental book The Mathematical Theory of Relativity, Ch.1.1. I concede that it is shocking and incredibly novel, yet I accept his ideas as true and will use them to indicate that time changed after Relativity and therefore will quote them again and again, for after all this is not a crisp, clean and perfect academic textbook, but a book of ideas about time that are so novel that many Publishers gave it a wide berth---they want books that can sell millions for them. There is no charity in publishing, but original ideas take ages and several interpretations and clarifications to reach the general buying public. Professor Eddington is right, and the fact we are not aware of the change in human conceptions of time is the fault of the shallow academics that followed him---more interested in making money through the media than proper scholarship; but if they're not serious thinkers how are they going to understand and support the serious thinkers like Eddington? I think the media moguls are destroying serious academic work.

---

[17] Thus we have it on the highest authority that space-time is true; Einstein was absolutely right. But in the Minkowski equation time is represented by 'i' then multiplied by 'ct'. This is mentioning time twice and it is wrong because time is not naturally present in space and you cannot use mathematics to make it so. Time is rather obtained from space with points. So the best formula remains the Einsteinian 3+1, as argued below. To reconstruct physics on the basis of equating space to time as Minkowski suggested is wrong. No one will even look at what I write about this issue let alone publish it, yet they are a thousand times wrong, equating space to time by mathematics is not possible, we construct time out of space. They say I should use mathematics to prove Minkowski wrong---and I have, namely his equation uses time twice as 'i' x ct, which, to me, is fraudulent.

Another categorical statement that I would like to add to Professor Eddington's missive is this: to my mind, till the end of life on this planet (not the religious 'End of Time' which implies that time is running through to a predetermined end), any proposition by whomsoever that opens or ends with the equation '$s=ct$...' is bound to be fatally flawed and must never be entertained, in so far as the 's' is for space and the 'ct' is obviously used to represent time. The logical reason is that such statements are always meant to show the equation of space to time or time to space; yet that miracle can never be feasible either in mathematics or physical thought, precisely because "There is no longer a universal time..."(Bertrand Russell). And if there is no universal time it means the universe has not time; for time is and can be created by anybody anywhere out of abstract and metrical space by the use of points as 'relation between points' (Russell again.) Therefore nothing, absolutely nothing, can be used to equate space to time as the time can only be had as a product of space as an independent phenomenon.

However, it is well known that scientists are not in the least concerned that the Minkowski proposal is not true of the real world. We all know that everybody describes it as 'artificial'. What the mathematicians wanted was a simple equation for space-time (like $s=ct$) and dispense with the 3+1 formula which is regarded, rightly, as less than objective in view of the fact that cosmic time is abandoned. And having got that from Minkowski, and further since time (as a way of getting required periodicities), is always the same, no one is in a hurry to join the philosophers in finding the best logical definition of time as both discrete and space-time---or 'discrete-space-time'. They prefer rather to glorify in the Minkowski creation of space-time by mathematics, no matter how flawed it might be in logic. Yet there are problems. One is that thinkers will never stop probing nature for logical truths. The reason is that the religions are always interfering with wildly imagined ideas of their own no matter how fantastic or even silly, and would nevertheless always gain mass following---not only unhelpful and annoying but dangerous. Secondly the problem of the passage of time remains unresolved, whereas by supposing that time is essentially discrete (since the year upon which it is based is determinate), the units of time in procession amount to the passage of time, simple. Time can thus be seen as logically traceable from physical premises as the product of space obtained with points and whose units in procession creates the illusion of the passage of time, and therefore time indeed does not flow all through the universe like some kind of 'A Gulf Stream Phenomenon'.

THE FOLLOWING IS THE NEW DEFINITION OF TIME FOR THE WHOLE UNIVERSE THAT WILL REMAIN VALID TILL THE END OF LIFE:[18] **every second is 'a space interval' and also, inevitably, 'a unit of time'[19], conceived (in the absence of**

---

[18] Also, see the Chapter on "The Order of Time Seen as a Matter of Arithmetic" below, especially the notes which define The Logic of Time in the Universe.

[19] This is where the conundrum begins---i.e. the lure to equate time to space. For how this happened is the very secret of how life came to be in existence, since the two are inseparable. However, it is acceptable by the 3+1 formula, but not in the Minkowski sense because the time, the very

a universal or cosmic time) as a moment of Being, Existence, the sense of Duration or the consciousness of the external world, mathematically constructed with points out of the space traversed by the earth round the sun in one orbit, as clearly mirrored by the clock's hand/s (whereby 31,536,000 seconds=one year). Without that there is no evidence of being alive. Thus, in conception, time and life (not time and space) are intertwined. Yet time is secondary because it takes a sentient being to construct it. But nobody can delay, stop or slow down the motions of the earth, so the time intervals or units move on unstoppably (that is one explanation of the oppression and urgency, Order or direction, the passage or arrow, of time: they're not physical but mathematical and conceptual). Repeat the process continuously and you get perpetual time as part of the logic of time in the universe. The problem is that it makes time necessarily discrete and thereby renders all other notions of time logically invalid, and some people resent that situation---especially the

---

idea of 'a waiting period' in nature, can only be obtained from space with mathematics, and infanticide is not known as an acceptable method in mathematics. I agree that it's a mathematical conundrum, but not humanly soluble---because the time itself is the product of space! What is certain is that it will remain the greatest conundrum of all in human thoughts. If time was running all through the universe in the form of a thread---general, fixed and absolute--- as previously supposed by religion in pre-relativity thought, then it could be feasible, but as the product of space, never. Ironically it was relativity that gave Minkowski the idea. Let me stress that it is time and life that are inseparable, not time and space. Rather time is the product of space, and life is the product of space and matter. In my opinion, time came after creation; it had no part in it, and it is inseparable from life because it is perceptual as well as conceptual, conceived out of percepts of perpetual motions.

academics trying to sell their tomes on traditional time. There is nothing wrong with the theory itself for that's how we've been constructing workable clocks for thousands of years, mistakenly thinking that the time was divine; the problem is with human emotions and religion now that divine or cosmic time is abolished by relativity.

ESSENTIAL DOCTRINES

What has been discussed above will now be stated in the form of irrefutable doctrines, but of course after fifty years of researching time in all of its aspects, in spite of the mockery, I have come to the conclusion that no matter what logic is produced and by whomsoever, there will always be writers including scientists who will argue for the religious view of time, all because people simply do not want to accept that death is the total end of human life. A large part of human knowledge is distorted or corrupted with human emotional beliefs. Yet, still, it came as a complete surprise to me that time is the most contentious subject under the sun passionately debated entirely in religious terms.

Nevertheless I insist on the doctrines. For I believe that all discussions of time should begin with its clear and objective definition because although we all use something we call time but define its nature and provenance differently due to our religious beliefs, leading to contradictory theories about nature and reality. Another reason is that science demands an objective definition it can use. For my part, I accept that after centuries of thinking about time logically, we've come to the conclusion that it is constructed by man as relation between points (Bertrand Russell); that it is also limited to a frame, so "there are as many times as there are frames", and cosmic or divine time cannot exist because the Lorentz discovery of local time means anybody can begin his or her own time from anywhere (Albert Einstein); and, therefore, time is metaphysically caused by the instantaneous spread, visual experience or any contact with the outside world (Professor A.N Whitehead.) Also Professor Sir Arthur Eddington (the founder of astrophysics), has observed that any contrary notion of time amounts to making "meaningless noises" as quoted above, since time is no longer seen as running smoothly ('even-flowing') through nature, and that, since this is the logical view, we humans have no other sensible or cogent explanation for time---and, quite rightly, as we have seen, he credited it to Albert Einstein. Professor Eddington's statement implied that he conceived time as discrete, not as an entity that is 'even flowing' through nature; and I have also found that it is discrete. And we recognize that Eddington was not just another scientist but the founder of astrophysics; he is also the one chosen by the scientific community of the whole world to attest the truth of the General Theory of Relativity. So we are in good company.

Therefore, as mere followers and supporters of these great thinkers, altogether, we now see time as the artificially generated periods by means of regular or repetitive motions for the total regulation of all activities and life generally, since it is 'constructed' from the general experience of the physical universe multiplied by motion or mathematically divided with points. Thus sentience, arithmetic, the ability to count and a theory of numbers were required. This shows that it was not an easy thing to acquire. Before that man lived by instinct and stimuli like an ordinary animal; language, ideas, thinking and planning came later with the rise of civilization. Of all the creations man has achieved, time is the most basic, most important and most mysterious---the foundation of all religious thought, even the means by which we question, amend and reconstruct nature. No wonder time has exercised human ingenuity for centuries, not least about its passage and continuity.

However, from the scientific point of view, the passage and continuity of time (constituting a single entity as 'time', always passing, never static)[20], is the least troublesome; we can even tap the finger to indicate the passage of time without arrows. For time itself does not move; only the cycles or motions we use to reckon time do move and deceptively seem to us as the passage of time, which, being arithmetical, can only advance through numbers---the years, for instance. The motions we use to reckon time are repetitive, which shows that it is not the motion of time at all. Time is 'Being'(existence); and we sentient beings while

---

[20] The seconds are moving on to minutes, minutes to hours, to days and so on to end in one year, then we start another year instantly. Yet discrete time as the mere awareness of existence cannot move; what we call 'the passage of time' are the cycles we use for reckoning time in units. For example, the passage of two years is just the passage of two orbits of the sun-- physical journeys. We call them years in the reckoning of our time in units. That's the nature of time: important only to human beings.

'existing' count repetitive cycles or motions to indicate the passage of time passing by our 'Beings', or how many cycles have passed by. It is a very complex and mysterious matter but human in origin. Take the years, for example: there are no 'years' in nature. There is only one year repeated to be years all the way to the centuries which are "**passing by**", say, the Alps, year after year after year. Mere physical cycles, not time, except that they introduce the concept of duration. It's not the Alps that are moving to provide the years but the cycles or repetitive motions we use to mark time. The mountains stay where they are forever as we continue to count the number of cycles that have passed it by and calling them 'years'. Thus any repetitive motions can be used to the same effect---tapping the figure is the same thing. Otherwise time does not exist as a physical entity. It's a complement to the human mind. That is why Russell deduced that we 'construct' our time---well, in the absence of a universal time our time had to come from somewhere, and we know that we are the ones who count the years to determine age. The factors for use in the construction of time may exist throughout the cosmos, but the time is constructed by us and does not, certainly, cannot be something running all through the cosmos and the same everywhere, since the parameters differ from place to place. Whatever 'Beings' are there in the cosmos would have to construct their own time systems by similar logical methods---this last point is my own personal belief as a certainty based on logical reasoning. The problem with logic is that it is science, meaning what we can trace or know in nature as of substance and/or really in existence in the cosmos. Like shadows, aura and influence, they may not be substantial, but they do certainly exist, or do occur.

Einstein often referred to 'logical thought', or 'scientific thought' and he was right. He and Aristotle (and probably Russell as well) were the most logical thinkers ever to grace this planet. Discoveries, inventions, literature, art and scholarship are necessary to create civilizations; but logical thinkers constitute the foundation of science.

As stated above, the culturally necessary units of time in the various lengths arise from the division of the earth's motions (and space traversed) by points into units or intervals, whereby the closely placed points provide short units and distant points give us long units of time like the year. So the seconds and years are the best confirmations for this theory. Time's divine status, arrows for its passage and absolute nature can all be consigned to the rubbish tip.

As defined in secular terms, time is now a branch of astrophysics; for so long as the year is our basic unit of time from which all other units are derived as fractions, time can be consistently deduced from experience thus: in any part of the universe any regular motions divided by points as required (into short or long intervals) will provide time units for cultural use as the logic of time in the universe. Time has no existence outside the human mind and what does not exist cannot be passing by. It is purely psychological counting of physical cycles as time units like the physical orbits of the sun as years.

One orbit of the sun is 'a year', a unit of time out of which all other units are derived as fractions thus: time in culture is known only in units. We pare down a year to the seconds and even lower. Each second is equivalent to about twenty miles in physical distance round the sun. At this rate, even a millionth of a second is quite a definite amount of space; so all of our time is derived from the year or the earth's orbit of the sun physically, which confirms the theory or notion that time is the equivalent of the secular coverage of physical distance, by which we are able to tell the difference between the lengths (or duration) of specific time units----that the hour is longer than a minute and so forth. Nothing is left of time to be explained with theoretical postulates as the universities, especially CUP, has been foisting on mankind to no avail. They still insist that Past, Present and Future cannot be wished away, and that even Einstein could not explain them to anybody's satisfaction. Yet he did. In logical thought, Past, Present and Future are truly illusions. The past is with us; it never went anywhere. It is what we carried with us to the present: our clothes, houses and other possessions from yesterday are what we are using today! For, after all, there are no days at all in nature. There is only one day. The temporary blips of the earth's revolutions across the sun do not change sunshine from day to day. The daylight, bar clouds, is on from the sun all the time. The earth's revolutions create the illusion of successive days, otherwise, from the sun's point of view, they do not exist. Thus the past does not still exist anywhere to be revisited by those religious theories about time travel.

The present is the past brought with us (we're still living in the same houses, for instance), and the future is unknown---which is what primitive man told us but we would not listen. It's now established beyond doubt, at least in my experience, that people just happen to be welded to ideas of reincarnation. To my mind it's as simple as that; and basic to that belief is that time is so mysterious that we can never know what it is. In fact, it is as defined above. Of course it is very closely associated with life and it is conceded that nobody knows why there is life; but time and life are two different things: time requires points, meaning that we had to come to be before knowing how to 'construct' out time. The long and short of which is that as the past does not persist anywhere to be revisited in the past, the future, too, is nowhere to be reached ahead of time (the time has not even been constructed yet); so the future is nowhere until it arrives through the earth's motions round the sun---and we know that it may never arrive. When people die (and we die all the time) their future is lost; it will never arrive, for everything is what it is only in human perspectives. That is why we cannot use animals, say lions, to do our thinking for us. While living as human beings we can of course use the imagination to speculate about the future; yet the imagination on its own is not reality, and yet outside reality there is only chaos not truth.

But we need to know the truth. If you take a million people with the freedom to speculate, you'll get more than a million viewpoints, as some people would have more than one opinion---we need truth to counteract that ruinous tendency in mankind. Man is powerful and at the same time highly vulnerable. Altogether, time is important and should be defined properly in logic. The best way to do so is, first, to abandon its divine suppositions, which we can now do reasonably confidently, due to the influence of Albert Einstein, as Sir Arthur Eddington has stated above in his definition of time. And it's about time people, including all those religious scientists, realized that Einstein's theory of time is far more important than all his theories put together, for after all we live by time, and we have to live. The strangest thing in the world, for me, is that his theory of time came out of a simple remark that the Lorentz local time idea can be defined as time 'pure and simple'.

As if that was not shocking enough, the greatest philosopher in the world at the time, our own (lovable) Bertrand Russell, asked the most important question ever posed about time by demanding to know what then is mechanized in the clock. I have since, in my own little way, been trying to answer Russell's query as detailed in this book. I think that ultimately time comes from nowhere and gets mechanized in the clock as nothing other than our habit of counting cycles and calling them 'time units' based on the period inherent in the earth's orbit of the sun pared down to the seconds. Everything in time is based on the earth's motions, for time is action and how long any action takes. Let me explain. Action is caused by energy, not time; so action in itself is not time or time controlled; it'd be either spontaneous or chemical; but time is then inferred from the action or contact. The time (or time normally), arises because of action, as we want to know the duration of any action; it is after the discovery of how to mark time that we can control activities by means of time for the purposes of creating civilization. So time was invented out of the conditions in the natural world and does not exist in the cosmos as a natural phenomenon.

Action takes time, meaning any action has to cover a period; this period is called 'duration', that is to say, 'during the period of the action'. But originally the concept of time was inferred from action---by counting cycles and calling them the number of cycles that any action took.[21] That is the mechanism the clock was invented to record----and can now be used to cover action through prediction and consequences after events, meaning planning purposes. All this has been explained in the book; without activity the need for the sense of time does not arise, reasoning and intelligence are also connected to the sense of time;[22] hence even animals do have the sense of time, albeit without the linguistic, mathematical and mechanical skills to mechanize it in the clock. By adding language, religion, fear of death, cultural practices, psychology and so forth to duration between events, time has become the most intricate subject on earth. But in the beginning although all the parameters for constructing time were there, man had no notion of mechanized time when we came down from the trees. As Russell has observed, time was a construction.

Finally, let me try to explain Professor Whitehead's highbrow or metaphysical definition of time---as I understand it (it's always a fallible attempt, for no one can claim to know anything for sure about time. All he or she

---

[21] Duration itself is caused, as explained below in the book. What causes duration is what we experience as time, and, although secular (mostly chemical), it may not be knowable in all cases. Thus aspects of time will remain mysterious forever.

[22] Existence without the sense of time is dour. Creativity, thought and civilization are all the consequences of having the sense of time. The reason 'Being' or life is so closely associated with time that they cannot be separated logically. The curse of religion prevented any thinker from discovering this idea of secular time until Einstein. Then God said 'let there be Russell and Whitehead to support him' and all was light!

has to do is to make any attempt logically valid). What he means, I think, is that time is an activity, vision or contact of any kind in nature; hence daylight can be regarded as the most basic form of human interaction with nature. Man's first attempt to reckon time was by merely marking the days as they passed and calling the process 'the passage of time' (I saw my grandmother doing that in the 1930s.) We still say the passage of time is exemplified by the passage of the days and nights---but there are no days; there is only one day, and it does not pass.[23] That is the quandary .If we analyze time very carefully we'll agree with Professor Whitehead that time does not move. It is rather like digital photography or images flickering successively one by one. But in real life, as the constituents of vision, the images are numerous and follow each other so closely that the effect is like moving pictures or images. When we add the counting of successive images that arise from repetitive cycles or motions to this digital images and call them the units of time, as we do in the reckoning of time, the total effect is

---

[23] Let me stress again that time does not pass because it does not move; what is moving is the repetitive cycle we count as the rates of passing time. It is an irony or a mystery, a paradox or sheer ignorance. Sunshine as daylight (or 'the day') neither moves nor passes by; it is on constantly. Rather we count the motions of the earth across the sun as 'days' all the way to the years, and pare the year down to our seconds and all other units of time. This is very important in the discussions about time. The main reason I am frustrated is that nobody understands this, yet it is true---there is only one day and it does not move. The concept of 'the passage of time' is completely mistaken. It occurs in culture but not in nature. All history (even the difference between yesterday and today), is the passage of events not time---we rely on memory and records.

the supposed passage of time. **The important thing is that the images do not stop, and also we're counting repetitive cycles as the units of time associated with the images at the same time. That's what creates the illusion of the physical in addition to the conceptual passage of time**. It is also important to realize that the units of time in the clock are parts of the yearly cycle and so can be regarded intellectually (that is, in abstraction) as 'cycles'.

Like Russell, there is no doubt that Professor Whitehead was extremely clever. He was also the first to note that the world of sense is a construction not an inference in so far as we can only see by means of light, yet light is also matter---in fact, the smallest bits of matter in our universe. So it means we construct images with the basic materials in existence. For man reality is limited to what light can reveal to him.

That is the definition of reality under the quantum theory---we construct reality with quanta, not infer it through the use of the mind, or even perceive the world on the principles of Plato's Idealism. Due to their professional vanity and inherent disrespect for philosophy, scientists shared the Nobel Prize for QED without mentioning Whitehead's theory, yet QED is the physical evidence that his suggestion was true---what we regard as reality of the external world is actually inferred from physical elements absorbed from the interactions between photons and electrons as Richard Feynman has shown. For once Idealism was logically refuted by G.E. Moore, there remained only one route for any meaningful philosophical reasoning. That's empiricism, but that doctrine ends in scientific enquiries and discoveries and is based ultimately on quantum electrodynamics, known as QED. Thus the Nobel for QED should properly have been shared between Whitehead, Russell and Feynman: Whitehead for the original theory, Russell for popularizing the idea in his book Our Knowledge of the External World linking the process to physics and Feynman for proving it so ingeniously in physics.[24]

After all, this theory, believe it or not, is the end of all human intellectual quests. The notion that light is reality, like Plato's Theory of Forms, proposes a solution of the ultimate query about human existence. It is the beginning and end of all philosophy and science, 'Being' and consciousness. But whereas Plato's theory has been refuted as not true of the world, reality as constructed with light energy is as solidly true as there is light, the same light by which we see the world. So in the end, the truth came as a very simple notion: what you see is what light shows there is, and the scholars who discovered this idea (Einstein, Russell, Whitehead and Feynman) deserve the highest honor. Individual academic, literary or cultural honors are not enough. We need a huge monument somewhere accessible so that people would ask 'what did they achieve?' and through that learn the greatest philosophical truth that settled the final conundrum about reality.

---

[24] The other scientific contributors were minor figures, but the Nobel committee seems to think that it is important to spread the awards, thus bringing nationalistic tendencies to bear, which I believe will eventually make them worthless. For instance, even though Einstein had been rewarded already, he and Bose deserved another. The awards should just follow merit.

The next quandary to confront this very clever American man---Professor Whitehead[25]--- was the problem of time because, as I said above, they are linked. He realized that to account for secular time under relativity some event (action, inaction or vision) has to occur; then counting cycles or applying units of repetitive cycles would give the time or duration of whatever had occurred.[26] Convoluted, perhaps? Yes it is. That's what time is, and the reason it's so mysterious---much easier to assume that God bestowed it on man and forget about the rest! By this theory Plato was not that clever, only a good writer; for Idealism makes thinking easy, but nothing in life is that easy, philosophically. Under the quantum theory, all knowledge cannot be mental because the light from anything can be blocked from reaching the human eye to result in seeing or knowing anything. Thus nothing without its light emissions can be seen or known. This is the final proof that all knowledge is not mental. Reality is out there and its lights must reach man for it to be seen or known. Idealism is dead, that is why the religion based on it is also dying---though only slowly, but it'll go.

For the logician the problem of time is to find an explanation or theory for the long period of the day or night, and that's when the counting of cycles came in. Any cycle will do, even tapping the figure is quite alright; but through human ingenuity we have evolved mathematical

---

[25] It was bound to happen that these four thinkers mentioned above would have something interesting to say about time, too.

[26] The sense of time is inferred from events---or even inaction.

techniques for paring the yearly cycle down to the seconds. So we say each episode lasts for many hours, or so many units of time.

Introduction

After the long Preface, an Introduction is still required because time is the most difficult subject in life---in a superficial sense it is life, or appears to be inseparable from life---and requires the apex of the human intellect to attempt a rational interpretation, however tentatively or even tremulously. If readers could be asked to be merciful, I would suggest that they judge this particular book not by the normal standards of academic books, because it is putting forward very, very difficult ideas about time, unavoidably using any method capable of helping readers to understand a theory which, if true, can help mankind to know time and life a tiny bit more accurately. One or two of the literary gurus who do not know the word mercy and treat every written piece as if it is competing with them for honors, have already classified my previous books as readable but some claim that it's only for the purpose of securing increased sales. How they know these things. So I cannot appeal to them for mercy. They will claim that this Introduction is even breaking a literary rule, as it is written in the format of a full chapter, but I cannot help it; it's all due to one clarification after another in the hope of revealing time to be what I think it really is. And, from my point of view, the only literary rule for a man of ideas is that his presentations should be readable.

Writing about time in philosophy is difficult because there is no method for separating time from life; and as the nature of life cannot be known by man except to live it, the true nature of time too cannot be known by any human being except to use it, and, through usage, employ logical principles to decipher it as much as possible---which is what I have been trying to do (rightly or wrongly) in so many books about the subject, on the presumption that in rational thought it is not possible to do more about any subject, least of all time.

The essential point is that time is known only through its usage. Even asking for the time casually means wanting it in association with something, activities or events: past, present or for the future. But this usage is the actual passage of time. Hence there is no problem with the passage of time requiring solution---for the passage of time is how we notice time; it is the time and we notice it through how it is used, and how it is used is all we can ever know of time. There is no paradox here; it is a simple fact in reality but never considered. Time is not a physical entity we can just grab and apply to events, or physically see it pass by. The timing is a conceptual interval in the mind linked to the physical space round the sun unstoppably moving on and on cyclically; so every action takes time (or uses a portion of this interval) because it involves a period in the mind physically linked to the space being traversed by the earth round the sun. In fact, it is a piece of the space round the sun produced by the earth. Every second is thus a portion of that space, followed by another so that time intervals are continuous and unstoppable---hence time is oppressive, not a second to waste because the earth's motions are unstoppable and so precise that 31,536,000 seconds equal exactly to one complete orbit for another orbit to begin, again unstoppably, and what applies to the second applies also to all units of time pro rata, including the year.

Again, let me explain time with another line of rational thought everybody can understand. We know time in usage only, but time is always passing; that means we can only know how it is passing and never what it is; it is nothing, completely insubstantial, just an idea in the mind and yet is in everything but no one knows what it is, only how it is used due to its being immaterial and yet oppressive as sketched above.[27] This is what has misled some thinkers

---

[27] It may be immaterial but the mind knows that time intervals are linked to the motions and environmental conditions on earth that cannot be violated without danger, since the brain is very protective. A living body is full of structured and intimately connected activities---a non-stop chemical factory. Time by the clock (meaning intervals between events that we've cleverly mechanized into the clock), constitute a signal that something is up, and the brain would spring into action. That's the

What is time?

who claim that time does not exist. Yet there is something we use as time, an unavoidable adjunct to all actions, except that our knowledge of it is how it is used only.[28]

I will hammer this idea home for misunderstandings can be costly; after all, originally the idea shocked me even after fifty years of thinking about time. What I have found is that we know time in usage only; every time it is mentioned it is in reference with something or some action; so if it is always passing whenever we notice it then we need no theories to account for the passage of time because what we know as time is how it is passing only and never what it really is. And I must stress that nobody knows or can ever

---

nature and purpose of time. Eventually it's grown to become a handy, complex mechanism for application to all events, past, present and future, but being buried deep in the brain, deciphering it is extremely difficult and thankless because people prefer the glib explanations.

[28] This is the reason time is inseparable from life, for every second of life is shown through action---breathing, standing, even sleeping, because when we sleep we breath and breathing is also a form of action regulated (or controlled) by time. Logically time is not inescapable in every action, but everything is interpreted in temporal terms due to the influence of religion which, in the past, used to be enforced with the law---even still now Sunday laws are used to enforce the holiness of Sundays in some countries. That's how stupid man can be. The real inescapable fact by which we should organize our lives is that in astronomy there are no Sundays---there are no days at all anywhere, only one constant day everywhere. The nights are caused by temporary shadows that do not affect reality in anyway, see below---e.g. do shadows change anything? Of course not. They pass by rapidly and leave nature exactly as it has always been: one constant day light everywhere.

know what time really is, since all we can ever learn of it is how it is used and passing by through the usage.

Of course, I argue in the book that time is caused by a variety of factors, including motion, chemistry etc, but really unnecessary (and probably impossible) to analyze it to the bottom---using something we can never define is scary enough![29] At least we now know how it passes by: it is the same as the usage. Like using soap for washing; how it diminishes is how it cleans, so when it is wasting away it

---

[29] The best logical definition of time equates it with being---aspects of existence mechanized in the clock as 'manageable units of existence', so that every second is a piece of your life, and also a piece of the space round the sun. We even put this in popular phrases like wasting portions of a woman's life (years) in a failed marriage, and so forth. But the religions got it wrong, saying time is what we are allowed 'to live'. Rather the life is physio-chemical and not predetermined in anyway. Time is our manageable pieces of the environment in which the life can flourish, as we cannot carry the whole existence (or environment) with us so as to be able to live: to say 'I give you ten minutes', is a metaphysical authorization given to somebody to live for ten minutes, or do something in that portion of existence which, it must be stressed, is going and will soon be gone and never return---so time is a metaphysical entity but conceptually created by man with mathematics. Though immaterial, a unit of time is a conceptual piece of nature---but physically substantiated by the space round the sun---for, after all, the whole of existence is conceptual and could not exist without the brain, thus consciousness and all existence evaporates when the brain dies and the body rots away. To probe time beyond this point will only end in nonsense. But of course existence, motion and activity on their own do not constitute time; time means that existence is mathematically parceled into pieces of intervals capable of being mechanized into a clock as I have explained in App. 1.

What is time?

means it is doing the job. Even if you merely ask for the time, some seconds are gone by merely uttering the words, because you used them as time in asking the question. This is the clearest demonstration of the notion that time's passage is the same as its usage and therefore there is no need for any new theory to account for the passage of time. As I have said, it means the passage of time was never a problem at all, except that time was wrongly conceived. To be honest, the idea was difficult to understand at first, but soon became as clear as daylight: the passage and knowing of time consist in its usage. How we acquire or invent the units of time is explained in App. 1 below.

In App. 1, time is quantified (that is, 'Being' or 'Existence' is reduced to a human language or to a human level as humanly recognizable units of being, for general application, though immaterial); otherwise the description given above makes it seem to be synonymous with 'Existence', 'Being' or a metaphysical entity that nobody can ever know what it is---which, in fact, is the case. Nobody can ever know the true nature of time. So time is a piece of existence. But when existence is quantified we reduce it, or being, to time units, as small parcels of reality we can manipulate to our advantage. Thus orbiting the sun is 'a metaphysical piece of being without human intervention', but when we quantify it we get our parent time unit called 'one year', which we then pare down to our SI unit of time---the second---and from the second or seconds (since they are continuous), to get all other units of time mathematically and culturally deduced without mythologies. Hence App. 1 is very important. It shows that to give your time is to give a piece of your life, and obviously it cannot be disputed that a piece of one's life is a metaphysical entity the nature of which is unknown---the same thing applies to time as a metaphysical entity or quantity known only through the mathematical manifestation of existence produced as a unit of time---from the year to the SI of time, the second, of which every other unit of time is a multiplication thereof, e.g. 31,536 000 of it amount exactly to one year; but it doesn't go on; instead we start another year, making the yearly unit determinate, as a result of which our time becomes discrete. Discrete time cannot march through the cosmos, therefore cosmic time and all religious time are concepts to deceive us. These sweet and logically consistent deductions about secular time began with Einstein and is part of the reason he is known as a philosopher/scientist.

But time does the job, so it is not correct to say time does not exist. By my way of thinking, time does exist because we're using it. However, it is utterly indefinable or demonstrable since it is conceptual (mysteriously inspirational, infinitely creative and always passing by), and not physical. For example when we reason that it is dangerous to go into the bush at night, or after certain hours, especially without light, we're using knowledge of time conceptually. That is why time cannot physically pass by; it is merely conceptual, but don't forget that ideas rule the world.

What is time?

We know time only through its intellectual usage---in other words, the brain is continually creating time sequences as we take physical action; but the time itself is not physical.[30] I concede that this is a mystery, in fact a conundrum, but one created by the human brain, the real creator of all mysteries on earth as in the universe. Indeed, age and ageing, distance, action, existence and space are all created by time in the brain: whether they are there or not only the brain can tell us; and what will happen when we're all dead we don't know---we can't even be sure they're always there as reported by the brain, neither can we know where the brain came from.

In any case, part of my argument is that the usage aspect of time can be proved, namely, any experiment will show that a person born blind who cannot engage in any activity at all will have no knowledge of or need for time all through life, no matter how long the person may live[31],

---

[30] Logically it can be explained as intervals between events or points that are bound to result from taking any continuous actions; and we all know that life is always in action, even breathing is action.. That is what I believe; Russell and Whitehead, Eddington and Einstein also believed it, but I don't know who else; and we know that arranged against us are the billions of human beings on the planet. So the problem of time is soluble; but to understand it one needs the combined brains of about ten professors of related subjects including mathematics, and how is that possible?

[31] The person (or animal) will no doubt have the sense of periodicity through personal motions, but not constructed, logical time linked to the external world and the universe.

simply because we know time through usage (usage implies activity), and the use of time is how it is passing by. So the well-known 'insoluble' problem of the passage of time never existed, except that time was wrongly conceived. Of course the blind person will remain ignorant of time sequences because time is a product of (or associated with) activity; so being unable to take any action at all will deprive a blind person of the sense of time. Secondly, ageing (growth, decay and any kind of motion, etc.) is either chemical or physio-chemistry and, as such, has nothing to do with time sequences; when they occur in nature without human intervention, they are accidental or random. Furthermore, our time system is discrete because it is based on the yearly cycle which is determinate, increasing unit by unit, or discretely---year by year as we know. Discrete time cannot march through nature, as Sir Arthur Eddington has emphasized above. All this has been discussed in the book. It is a serious matter but because it is about time---and everybody still believes that time is eternally bestowed from above---I have had to rely on the POD method of publication amid mockery, or as if I have never existed. Now somebody is asking the whole world through the internet to ignore my work because my country of origin is not noted to be capable of producing philosophers---a matter for the anti-racism bodies to deal with!

    Still, let me state the facts as clearly as possible, or as I am capable of: we can know time in usage only; and in usage it causes (or creates) all existence or being through regulated phenomena. Hence if time is now seen as something created by man, then the search for God should be centered on the nature of the human brain which creates everything human. This is to recognize that the search for God is assumed to answer all human problems. But what all thinkers down the ages missed is the fact that time is a human creation as Einstein and Russell have now established. My little contribution is to point out that the brain constructs time and time regulates life, therefore it is the nature of the human brain we need to investigate.

Of course, it's easy for the apostles of revelation, who receive their knowledge through the bed room window when they go to sleep, to tell us what it is without evidence, but in rational thought we know time as the biggest conundrum of all, yet a magnet to accursed souls with a thirst for knowledge who would search for the truth no matter where it led them. In this Introduction, I shall repeat again and again the scientific or logical notion of time upon which the book is based (because the religious people are stubborn), namely, it is acknowledged that the sense of time (or natural time) is the sensation of interval between points---also known as a period of waiting, or relation between points in Russell's phrase.[32] I believe this

---

[32] As a sensation it means time is altogether human and the foundation of knowledge, since it becomes the source of cataloguing objects in physical reality and thinking about 'them'--- the basis of the Subject/Predicate axiom, and probably the reason time is so intimately bound-up with life and the workings of the mind. Personally I have no doubt in my mind that time is the basis of thought; therefore any

is what inspired Professor Whitehead's definition of time as 'non-interacting moments', the moments being contacts or sensations. They're non-interacting because that is the true nature of all sensations, contacts and perceptions: the origin of individuation, which is a very important issue in philosophy.[33] To construct the clock for the general use of this facility is the human contribution to the having of time.

The essential point for the interpretation of time is that without the human creation---or construction---the mere intervals between various events occurring in the wild even among the beasts of the forests could not be harnessed for the benefit of civilization as evidence of time going, or as time for short.[34] Otherwise all events occur for their own purposes, not the specific purpose of recording time. In rational philosophy it is assumed as self-evident that, apart from 'physical causality', there can be no conscious direction of events outside human society. For example, several events (perhaps billions) can occur simultaneously; and how can we use all of them to reckon time? Therefore we use only one, the motion of the earth, to reckon time in

---

'Beings' in the cosmos would have different time systems and technologies, and therefore probably dangerous to us. Everything in the universe is unique, so is man and his time.

[33] Professor Whitehead was very clever indeed. He's a mathematician and mathematics is vital for the study of time.

[34] Rather it is after the construction of the time system that we can interpret any motions and events in terms of the units of time already established---i.e. as to whether they are seconds, sub-seconds, minutes, hours and so forth. The days do not count because there is only one day.

What is time?

order to cover everybody on earth. The result is that billions of other events do not count; even the rotations of the earth are not relevant. Thus the Day&Night system is metaphysically useless for the interpretation of time, since it is a temporary interruption of reality---otherwise all reality is daylight. In 'our' universe we <u>know</u> only of daylight everywhere, or constant daylight, therefore there is only one day, and 'the days' are cultural inventions that do not exist in the universe as distinct entities or phenomena. This is not unique; we have numerous cultural inventions of no metaphysical significance, the most important of which is language. In metaphysics language, like time, means nothing to anybody except the local users. H.A. Lorentz had no idea how right he was to call his discovery 'local time'.[35] This innocent discovery about time (which he failed to realize was important), is the single, most profound intellectual achievement in human history, more important than anything in life because time is essentially life, and it

---

[35] It appears that the chance for the existence of life was a narrow unplanned chink (error or accident) in material reality, and it has no significance at all; we are, I am afraid, orphaned at birth and it shows because we die soon after birth if not taken good care of by our parents. Nature makes no provisions for the survival of infants; it implies that nature knows not nor planned births---it's entirely random and arbitrary. The real mystery is the brain which keeps us alive for some time before coming to grief, something like a computer coming to the end of its life---and electronically it's quite possible for a computerized entity to emerge from all this fantastic flux in the material world that are mostly powered by electrons and photons acting together, as they do from the eye to the brain whence all actions are engineered and executed.

led to the view that time was not given with religious implications but created by man to begin the process of demystifying time, life and religion. Of course, man still has no idea of the origin of life, but all the changes in our theories in science and philosophy, beginning from Einstein's unique proposals, are based on the concept of time (all time) as being 'local time', not general, absolute or fixed, but locally created for local purposes, and not bestowed by God either, as He doesn't live with us locally! So the one essential element in life (the time for living) at once became rationally deducible without mythologies. It also became clear that human beings did not appreciate truth unless it promised glory, everlasting life, power or wealth.

Yet, obviously, showing that time can begin from anywhere was not only important but also introduced (or created) the tendency to base all human thought on the rules of logic, and the whole of philosophy (done properly as rational philosophy) became the clarification of the rules for logical thought. **The reason is that time is obviously synonymous with existence in psychology. What is contentious is how it passes by physically, and that brings in mathematics, and since mathematics is part of the scientific mode of thought, the whole process renders science supreme, as Bertrand Russell used to say. The supremacy of rational or scientific thought opened up society and depressed all religious mythologies. The apparently simple observation that time is not fixed**

changed the intellectual landscape of mankind. For time is not only mysterious; it controls or underlies everything.

This is the history of how science came to dominate human thought, and it's very recent, mainly (even solely) due to Albert Einstein's notion of time. It changed the world by making time phenomenally important but secular, thus putting Heaven and Hell to permanent sleep. Yet, strangely, the academics who are supposed to know how to reason rationally do not like it; they do not regard my work, for instance, as useful to philosophers for failing to indulge in the Wittgenstein brand of meaningless verbiage; I also accuse them of being shallow for failing to note that after Russell and Einstein philosophy has changed from mere arguments to the logical clarification of the rules of rational thought and guidance to science. On the other hand, the ghost of Wittgenstein and the current scholars of Oxbridge (the brains of Britain), who call him the greatest thinker of all time---because they're brainy---struggle to advance arguments to make even physics seem worthless, and yet remain the most avid consumers of the products of science.

A recent headline in the papers that caught my eye states: "WORLD NEEDS MORE SCIENTISTS OR IT WILL RUN OUT OF IDEAS"[36], yet the scientists also need philosophical clarifications in many sectors of their fields, not mere verbiage from Wittgenstein---why didn't Russell mention him even once in his monumental History of Philosophy?

---

[36] The "London Times", Saturday, 24th Dec. 2016, p.30.

He said later in My Philosophical Development that it's because of his logical mysticism, which was very seductive and capable of making you commit suicide, but stripped of the linguistic verbiage, means absolutely nothing; and it's not philosophical, especially after Einstein and the quantum theory. But are we being invited to conclude that Oxbridge no longer has the capacity to judge such works correctly but rather praise them?

"We used to have good-quality books, music, television and cinema. Now we have rubbish across the board"[37], said the comedian, Alan Davies, and quoted by the Sunday Times, 1st Jan. 2017, P. 28. We have all noticed the intellectual decline now afflicting the whole world, and soon such comments will overwhelm us, before we realize that electronics, the computer and internet are not making us smarter yet we create everything. It implies that we're subsequently creating less efficient machines than would be possible if we're getting smarter by the previous ones— in a word, we're growing dumber, not brighter, by (present) technology. That's nobody's fault; it is rather a dilemma. Innocently I call it "The law governing the frontiers to our scientific achievements"---namely, barriers to further automation and scientific development may come from mental decline caused by the benefits of existing automation and artificial intelligence, since we create

---

[37] Comedians are artists, and artists are pretty good observers of social patterns, the reason they can invoke appropriate reactions from people---make us laugh, cry or angry like nobody else.

everything by the same brains. In other words, since we're using the same brains, if we're rendered mentally lazy, complacent or self-satisfied by the success of current developments, we could scarcely advance further in the future unless we grow new brains.

We invent technologies and rely on them completely to undertake tasks for us so that we can rest our brains; that, we think, is progress, scientific progress---we're reaching to the moons and planets. As we do so we become mentally less and less efficient, leading to eventual collapse. That's my prediction; it may not happen today or tomorrow, but that's the way we're heading. I even believe the first signs are beginning to emerge; artists have begun to notice them. Anybody who believes that our technological advance will continue indefinitely is deluding himself. There is bound to be a built-in or random obsolescence; all nature is like that, random, chemical or accidental; chemical causes give us the false impression of planned action but there is no good philosophy to justify that. Life and death appear as a pointless repetitive cycle. That is what Pythagoras and his people misinterpreted as 'rebirths'. It's not rebirths, but a continuous cycle of life and destruction, a cycle of dying and dying, bearing and bearing; something will always come to replace the last thing as it is destroyed (even fire leaves ash behind), but what it will be is never known, and rarely predictable through experience, due to the infinite creativity of the cosmos. Some pretty clever artists have imagined this---i.e. by sketching Extraterrestrials in many different forms. They are right not to draw them like human beings. It is most unlikely they'll be like us. The universe can create infinite varieties of everything all through random action. The only permanent things are light, motion and the activities associated with or caused by them---plus chemistry and physical contacts.

Everybody knows that death is not the end of society, that's the justification for making a will. Yet we know that life on earth will disappear one day and everybody will be dead upon the demise of the sun; that being so, can anybody imagine what will happen to the universe? Will it continue to be there, and if so in what state when there are no human beings studying its comings and goings---and will such activities be going on at all? I am surprised the poets and philosophers have not subjected this idea to deep thought. For, plainly, the greatest mystery is the brain's capacity for probing nature. When it is gone who is going to imagine what---and with which organ? Well, from my point of view, there will be no time or regulated life for a start. Of course, at least one of the factors we use for the construction of time sequences---the regular motions---will be there, but who would be counting the orbits of the sun as years and pare the year down to the seconds as our SI of time? So I have concluded that this is not a topic for profitable discourse. The meaning of life to man, and even the meaning of the whole universe once born into it without your consent, is to obey the law and live peacefully, enjoying what little comforts come your way, do good for good to follow you, look after your health and try

What is time?

to leave something to your descendants and others in society. Any religion that fails to teach this sermon is failing its flock.

It is most likely that we are unique and yet scarcely important---important to what or who? That's the most logical explanation to account (even tentatively) for the infinite vastness and complexity of the universe. Everything we see or know about the universe is bound to be variable or contradicted by something perceived or imagined, a boon to priests and charlatans. Yet even the imagination and all our fantasies are part of the contents of the universe, something existing somewhere, if not physically then it is in the brain!

Anyway, I may be regarded as old fashioned[38], but I've always thought nobody in his right mind would dare to contradict Russell without a strong dose of advanced logic, yet, typical of this country, soon after his death he was treated like an ordinary writer, with somebody (a cheeky and ungrateful foreign import) calling him 'a mere scientific populariser...'[39] In fact he was the greatest philosopher

---

[38] I'm like an ancient fossil and very ill. By the time this book comes out---only through POD as no publisher would invest in me---I'd be nearly 80. The advantage is the fearlessness of human fossils---dead sheep fearing no knife.

[39] Because he condemned Wittgenstein, who turned out to be much less of a philosopher, while Einstein was showing that philosophical ideas can change the world of sense and therefore no longer a joke played by under-developed, childish minds. In all history nothing so amazes me as

since Rene Descartes, and probably as clever as Aristotle. Everything I have said in my ten books on Time can be inferred from the works of Bertrand Russell[40], even before we come to logic, the rise of science, 'Our Knowledge of the External World', the mathematics, the histories, ethics and politics. Quick as a flash, he realized before everybody that if time is not cosmic, general or absolute then logically it must be a human construction to open the gate to secular time propositions that are scientifically deduced from perceived phenomena, as I have been trying to show in my ten books on time with all the mockery heaped on me by some of these foreign tramps. Yet even in this Introduction alone any good thinker can see that secular time has more to recommend it than the old idea of a universal, absolute time programmed to march straight to Heaven and Hell, which even school children will now be able to condemn as plainly false. It is now common knowledge that, according

---

much as the appointment of Wittgenstein to a Chair in philosophy at Cambridge; but then the university is something like a theological college redeemed only by scientific thinkers not the rest who go there to squander money from wealthy parents on booze and socialize for the lifetime network of influence. The all-important thing is the name. Oxford too would be worst than a Convent except that it is redeemed by the OED project, and when the docile civil servants distribute honors to the media cardboard celebrities, they forget that every year somebody from OED should receive a gong.

[40] They can all be taken as comments on his observation that time is 'a construction', in his book Mysticism and Logic, or 'What is measured by the clock?' in ABC of Relativity, but I cannot expect the charlatans to understand such issues.

What is time?

to relativity, time can begin from anywhere and, being discrete due to the determinate cycles used to reckon time, it cannot march. The idea that every time is local time for local application, there and then, is the only rational explanation for time and we owe it to Russell, Whitehead, Einstein and Lorentz.

However, it's about time the British learnt to look after their native thinkers more than the cheap and offensive, almost criminal, foreign imports who have offended everybody in their own countries and are just running here for sanctuary without gratitude. In my country, which one of the foreign tramps claim is not known to be producing philosophers, we have a primitive African proverb which says: salt does not have to proclaim that it is sweet. We cannot do that for this country but, let's face it, before Einstein shocked the world, Isaac Newton of this tiny Island led science as the most profound scientific thinker of all time. We also had George Stephenson, Faraday, Darwin, Alexander Fleming; even parts of Einstein's greatest theories came straight from the Maxwell equations, etc., etc. The list seems almost endless. Yet like salt we will not claim that we are palatable---but that does not mean any tramp given sanctuary here can insult our greatest thinker cruelly because he is no longer here to defend himself.

There are even more serious problems with the British and the rest of the world at the moment, namely, the inability to assess anything without media money, media hype, media treachery and the internet lunacy associated with them; intellectual laziness, over-reliance on electronic solutions, the dying art of thinking and writing clearly, especially by the young, and the intellectual decline of our dons since the great days when the quantum theory was being developed. Meanwhile fantasy peddlers are making billions changing the meaning of value in literature. Of course mentally lazy children would love magical, fantasy tales not mathematics. All the time we hear that Wittgenstein was either a great thinker or one of the greatest, and the study of black holes deserves a Nobel---for what relevance to life on earth, I wonder.

Moreover, as I see it and never tire of pointing out, a great deal of the problems we encounter in physics and cosmology comes from the traditional philosopher's mistake of regarding time (still) as a natural part of physical reality[41]---especially in cosmology---rather than an artificial contraption created by man. In fact, what seems to be time anywhere outside the earth is caused by either chemistry or random action without conscious direction; even the very notion of time resulting from a period of waiting is human in origin because it relies on the intellectual use of points. As explained in the book, the human mind is conditioned by our time and we carry it to apply to other

---

[41] This is a religious idea yet everybody believes it, hence my proposals are not even read at all. Yet as Sir Arthur Eddington has pointed out, because of Einstein's researches, time is no longer seen as flowing through nature, or naturally existing, and certainly not divine; many of its constituent elements are known to exist everywhere, but it takes the human brain to create it. Most of all, it is discrete and does not flow through nature like a stream.

worlds in contravention of the Einstein theory of frames, because the cosmos is governed by chance and not by regulated time such as we have created. We have to remember that we did not have it before; it took us millions of years to invent time. Never mind that the elements are available everywhere---the maturity was not there, since it requires the intellectual use of points as I have said.

This, I submit, is the closest we can get to making sense of the cosmos. The cosmologists claiming that they've discovered something interesting about time in black holes are sadly mistaken if Eddington and Einstein were right, and, at my age, not being afraid of the sack, I have made that plain in the text below. There is very little of value in the study of black holes, except that nowadays Britain will always call any scientists of note 'as brilliant as Einstein', and shower him or her with so much honors that the person ceases to be capable of any more serious work. It's all the fault of Einstein; he's so good and so extraordinary and still so simple that every country thinks they can produce somebody like him---not a chance. Being a thinker, I know how hard he must have worked and also how immensely efficient and unique his brains were. It's unlikely we'll see another Einstein for millions of years. The best we can produce would be light years away. Something 'universally' unique happened to Einstein's senses for logical thought. We don't even completely understand relativity and the quantum theory as yet! The metaphysics underlying the equation of mass to energy is also mysterious, as it makes reality questionable in whatever form it is encountered; it is even scary when considered in the light of the Higgs bosons. Then there is the problem with the speed and nature of light, the dark matter universes as against out light matter worlds, interstellar distances as measured in years and the temporal length of the year itself under scrutiny as our ultimate yardstick of time etc., etc. All this on the understanding that time is not cosmic or divine---and we owe them all to one man, Albert Einstein.

So then, the days (as already mentioned) do not exist naturally in nature. I repeat, in this universe, all physical reality is permanently illuminated; the night-times are freakish, temporary and unimportant, they cannot change anything in physics. Everything remains the same because the night periods are rapidly passing shadows and in no way important in material terms.[42] There is nothing we can only do by night and never by day. Again, the nights are not known in astronomy because they are flippant, quirky shadows of no significance.[43] Of course they are important in human psychology, but how can the conditions in human psychology affect physical reality? We think we're important because the Oxbridge dons say so to get their research funds increased, yet we are absolutely nothing,

---

[42] It may be true that there are light and dark universes; if so then we live in the daylight part of the cosmos.

[43] Imagine trying to interpret the world through the shadows of a huge bird's wings blocking the sunlight---yet the nights are not different in kind only in degrees.

and how we came to exist for this brief period is the eternal, unfair mystery---to suffer and die off without recompense. It's good that life does not continue after death.[44] Can man use the mind to avert natural disasters like earthquakes? Or can we survive the demise of the sun as species? There are billions of events like the Day & Night system everywhere (footsteps, clouds, rain, snow, shrubs and how billions of leaves wave in the air, the billions of noises made by all sorts of objects, animals and trees, etc.) that are literally countless, quirky and meaningless in a universe of this size and complexity. In fact, time is one of them because the universe is not regulated by time---which is dependent on humanly created points---but by random events, albeit with a certain logic of their own caused by the effects of gravity and chemistry, motion and accidents. Time is required in constructive actions; but the universe exists on random activities, hence cosmology can never be an exact science. There is causality but only by chance.

Similarly, ordinary events and motions do not show or demonstrate the passage of time---simply because we are not using them to reckon time, and Russell was right to call time 'a construction', an idea without which I could not write this book. But the creation of time by man in this manner to link it with physical reality, motion and

---

[44] Those of us praying for resurrection, reincarnation, rebirth, hereafter, life after death etc., do not know what they're talking about! The world does not bend to human will. To my mind, a painless death as the total end of life is a personal blessing, certainly preferable to a life of misery.

astronomy was so complex that it must be accepted as the greatest philosophical achievement of the human mind. No wonder time appears supernatural to simple minds. In fact, future developments in science could show that the creation of time was nothing special---just the natural consequence of living and perceiving or sensing reality. The real, everlasting mystery is the nature of the human brain; it is an everlasting conundrum because that is what we use to think, and nothing can explain itself in metaphysics, meaning the ultimate reason for its existence. So we cannot use the human brain to explain how it came about.

Meanwhile, the origin of time in logical thought is not known either in science or philosophy, other than the retort that 'it just is', almost as senseless as the religious assumption that it is a gift from God. Writing so seriously about it has exposed me to derision; nobody will even agree to read my manuscripts because (somebody says) my country of origin has never produced any philosophers. Altogether not a helpful way to discuss an aspect of life so vital to everything we do, and yet the religions claim that it is fixed, absolute and the same everywhere so that a second here is a second all over the cosmos which, obviously, cannot be true, though people believe and organize their lives by such dogma helplessly. It is the duty of logical thinkers to come to their aid, since logic is the ultimate arbiter of what is true and in real existence that human beings can rely on in safety---the main purpose of science and philosophy.

What is time?

Actually a rational review of the nature of time became necessary because (as stated by Sir Arthur Eddington, no less), Einstein showed by his researches that time is neither absolute nor fixed but rather varies from place to place due to H.A Lorentz's discovery of t_1, also known as local time. Not long after that Bertrand Russell asked the most important question ever asked about the nature and provenance of time: "If cosmic time is abandoned, what really is measured by a clock?" (ABC of Relativity, 1925, Ch.4.) Before that, in about 1917, he also gave the world this ingenious deduction as the greatest philosopher alive: "It seems that the all-embracing time is a construction like the all-embracing space. Physics itself has become conscious of this fact through the discussions connected with relativity."---Mysticism & Logic, Ch. Viii (x). I found this extremely helpful. These are serious questions, yet even Einstein had nothing to say about them in physical theory, no doubt because it is a metaphysical question, and helps to show how metaphysics is vitally important in human affairs, I mean even in science thought.

Einstein's Two Postulates of Special Relativity failed to define time for science use, even if not for the rest of us. Without doubt he was extremely good, incomparable, the greatest in history. To me he was greater even than Aristotle, not because of recent academic confirmations of general relativity---they did that only to get increased funds for their work. Otherwise General Relativity was confirmed a hundred years ago by Eddington. In science, because of money and awards, many academics are fraudulent. But I think we must now write time as the third postulate, namely that any frame (or planet) to be separate and independently self-sufficient, must have the necessary parameters for the creation (or 'construction' according to Russell) of time since it is not generally permeating the cosmos as a providential bounty---fixed and absolute---yet vital to life.[45]

To his credit, Einstein said "There are as many times as there are inertial frames", which implies that there is no time anywhere until any frame creates its own time. By implication there can be no time in any part of the cosmos where no sentient beings had created or constructed their own time system. This is what I call "The logic of Time in the Universe".

From my point of view, time imposes order, it is a period of waiting and the units (like the years) are relations between points (another Russellian phrase.) Other units of time are derived by paring a year down to our SI unit of time, the second---and this includes the atomic time because it has always to be related to the second to make

---

[45] The importance of this is to impose a minimum size for 'Bodies' that can be called 'Inertial Frames' where life can flourish, in that they have to be capable of their own independent orbits of the sun to provide the necessary 'base unit' for the logical construction, analysis and meaning of time. The reason is that we know only of the system of time where all other time units are fractions of the yearly cycle, so that they can be related to the space traversed by the earth round the sun to mean something to us. Whatever we do the sun will always be the source of life, time and our principal source of light.

sense. All this is possible only under relativity. The academics have failed to explain that the Einstein notion of time (as limited to a frame), is far more important than his theory of gravity. For if it is limited to a frame then how does any frame invent its time---and also what is time in the cosmos? In other words, does the cosmos have time?

That question is metaphysical, and when it came to the metaphysical status of time, Professor A.N. Whitehead was at hand: A time system, he said, as intimated above, "is a sequence of non-interacting moments [however that moment is defined---a year is one such moment because it is determinate and has to be repeated to continue together with all of its fractions.]" And deriving from this, "a moment of time is to be identified with an instantaneous spread of the apparent world..." (The Principle of Relativity, Cambridge, 1922.) This means any human contact with nature, and I think it makes our time necessarily discrete so that, for a start, time travel becomes a myth.

Discrete time cannot move; the succession of the units or separate moments (like the years) creates the impression of the passage of time. Let me stress this point well: extremely fast succession of images creates the 'steady' objects and actions we see in cinematographs. Similarly, the succession of time units (as metaphysical contacts with nature) creates the illusion of time 'moving on', or as 'the passage of time'. In reality the passage of time is the passing of time units 'in procession', not as some kind of a stream physically passing by and since that is all we can ever know of time (because time never stands still) we are deluded in thinking that time is passing by[46]---- however it means the passage of time is the time, or what we know and use as 'time'; all time is always passing.[47] It seems ironic but true, and certainly not a paradox. What we call time is the passage of time only and never the real thing. Real time is unknown; whatever it is, only how it passes by can ever be known. But I suspect time is caused by certain factors, agents or parameters in our environment, and certainly not divine. Not divine but so complicated that I believe nobody can ever discover what really causes time. But how it passes by, I hope, is no longer mysterious, though not like a physical entity in a stream, but unit-by-unit in procession. The truth is that the passage of the images (contacts, percepts, etc,) occur with such speed and are so small due to the invisible nature of atoms that the passage of time seems to be something like a stream; however, since even atoms are individual, we have to accept that the passage of time happens as the succession of time in units and therefore discrete.

---

[46] All we can know of time is how it is passing by because it never stands still---yet, ironically, it does not even pass at all; the units of time are in procession, not physically passing by. You have to believe you are insane to say this, yet it's true: as a discrete entity rather than a thread, time does not pass through nature. The old religious idea of time is completely mistaken. I having great difficulties in this career of mine, but that is right. A fundamental quantity like time cannot be changed overnight.

[47] It is all very confusing, because we have to use repetitive cycles to track or reckon time; so there is motion at every turn and we assume it is the motion of time, but that is wrong.

What is time?

This idea requires more logical explanations. The book is full of repetitions for fear of being misunderstood---since time is such a difficult subject. How do we know that time is passing, and where do the units of time come from? My answers are these: It is a mistake made thousands of years ago that the passage of time is indicated by the passage of the days and months and years upon which the clock is based. Man had nothing else to cite. In fact, there are no days, and the months and other units of time are fractions of the year, which is also only one and has to be repeated to become 'the years'. So the years do not show how time is passing by because they do not, in fact, pass at all. There are no years to pass by. There is only one year, repeated over and over again. Thus the individual years are in procession: the procession of successive units of time creates the false impression of 'the passage of time'. But to know this we had to realize, as Einstein was the first to do, that time is not providential, fixed or absolute, as was suppose in traditional time out of ignorance.

From this simple idea, the problem of the passage of time is solved. For science, philosophy and any person interested in rational thought: there is no physical entity passing by as time; it is purely mental calculations for cultural convenience. Likewise, history is the march of events not time as the time is associated with the events showing 'when' they occurred. For instance, Journalists report events and when they occur. In all cases of antecedents and consequences, only events are involved. Time has no before and after, only events do. Looking at time strictly logically, this is what we find as scientifically acceptable.

Technically, without going through the complicated mathematics of Hermann Minkowski, let me point out that discrete time means time is not running all through the cosmos like a thread so as to curve together with space to make time travel feasible. The equation "S= CT...", which became fashionable as a result of the Minkowski ict equation, is logically and mathematically flawed---like the whole Minkowski proposal, which is mistakenly used simply because mathematicians have fallen in love with it, not because it is materially true. It's not true that what affects space affects time as well automatically because they're not connected. The only logically feasible merging of space and time is by the 3+1 formula. Space and time were made separate in special relativity and they remain so to this day; it means the 3+1 system with the proviso that they are literally inseparable, although the time is 'constructed' by man. Inevitably this brings in the biggest metaphysical conundrum in human life, namely, did man begin life without time and yet created it to become so intimately associated with life that they have to be quoted together at all times as space-time? So far the scientific answer is yes, except that what affects life does not necessarily affect time as well without complicated mathematics---but there is nothing we can do about it. It may imply that, because of time, reality is not completely objective, yet it works; all physics is mere chemistry and chemistry will work on the notion of "a period of waiting", which is the essence of

What is time?

time, no matter how it is recorded, called or came about. I personally believe time will forever remain mysterious, but we've come close, very close to its true nature---as a period of waiting in the mind.

Of course life is in a mess. We don't even know what time is; but we do know that the clock is indispensable and it makes time discrete. Since we know how the clock came about, it is easy to trace secular time to its historical foundations----and we find it to be discrete, i.e. based on the earth's motions. Mathematically, because time is "constructed" by man under relativity, I repeat, it can be linked to space but only by the 3+1 formula.

I believe Whitehead is right, spot on----very clever indeed. Let me explain. Time reckoning arises solely upon contact, perception or activity of any kind in the world; otherwise if a person is blind, sits still and is insensitive to motion at all, he'd have no sense or need of time---but that is dead life. For normal life, time arose because of activity---contact with nature (any moment) as Whitehead put it. However only activities in social settings matter in the interpretation of time; otherwise there are activities among the beasts, but without civilization, and man's searching brains, they do not add up to theories in metaphysics.

One logical consequence of this understanding is that the universe itself has no time. What may look like time intervals to the human eye in the cosmos at large is mere physical or organic chemistry, accidents, natural processing or gravity, delayed reactions or inertia; above all, the human eye is accustomed to time sequences and tend to regard every gap as 'time', every motion as 'time going' and the Day&Night episodes as the march of time in the cosmos. This is a mistake in the interpretation of phenomena. "Moving on" is also interpreted as "The march of time" by the religions and mathematicians---who regard themselves as partly religious---and this concept of 'moving on' is essentially the counting of successive days. Yet there are no days! In fact, moving on is the march of events not time. Examined logically, time does not move. All history is the continuing story of what incidents have been happening to man from the distant past and still going on; events do not stand still, they move on. The times of events are the stages in the orbits of the sun at which points the events occurred. For the orbit of the sun has been divided into individual seconds ticking away in the clock, so that every point in the orbit is a unit of time. Indeed history has always been a source of worry and doubts about every theory of time. Let us examine it in a little detail.

Evidently the story of the past appears to be moving on or marching through the cosmos from the distant past to tomorrow and infinite future in time. There was the first century to today, and we are still accumulating the years towards more centuries in the future. We can't blame those thinkers who believe that time (as exemplified by the march of history from the past to tomorrow) is marching pass by; and as the religious thinkers believe it is a Biblical movement towards the Day of Judgment, scientific minded writers are either lost for words, or mumble something about endless time---or just say anything. In reality there is

What is time?

no yesterday and no tomorrow either. We continue to live in the same place and everything is the same, give or take a few incidents or events. Of course the earth keeps turning round creating the nights, but that has nothing to do with either science or philosophy. Walking in the sunshine also creates shadows, and we don't change philosophical theories because of that. The day and night events are clearly caused by the temporary rotations of the earth; that is all there is to it. In the entire universe everywhere there is only one constant day; the rest are temporary events across the face of the sun; they have no metaphysical significance either in science or philosophy.[48] Using them to interpret the nature of time (which has this insurmountable, fundamental significance to life), is

---

[48] Time is a fundamental, necessary and indispensable accoutrement to life; gaps between events, relations between events or intervals between events which we know as 'time' occur in every action we take; as such it is metaphysically important, vital to human existence; but the erratic motions of bodies are obviously not part of the fundamental requirements of existence, simply because they are erratic, unplanned, unpredictable and infinitely variable. So the rotations of the earth which result in the days and nights, weeks and months, in our culture cannot be natural elements in the interpretation of time---like one person's erratic habits, how can we use them to interpret reality? In other words, we use such quirky things for our own convenience, which convenience is of no consequence or relevance to the cosmos in any way. Again and again we need to be reminded that man and all his values are absolutely worthless to the inanimate cosmos. I agree that the situation gives kudos to the religions, yet at the end of the day, man has a hard life to live, and only evidence-based ideas are useful for that, so, inevitably, the religions and their myths get relegated.

absolute farce. What is wrong with the honest admission that we simply do not know the true nature of time; that, for instance, the year or any unit of time can never be defined logically without reference to other indefinable units: 365 days give one year, but how long is the year as a logical unit without reference to days? Thus I argue that using the year to estimate the age of the universe is also problematic to say the least, and even then such ideas have no educational values.

The religions are partly to blame, and they have never been good thinkers. They want to change the world out of their own desires, weaknesses, disadvantages and sins. The day and night system is nothing important, except that it warns us not to go into the bush at certain periods. Philosophically it's just like a huge bird flying over to block the sun for a moment or so; and we give a name to each moment the sun is absent. Whatever may be the duration of such events they do not create past, present and future as part of physical reality---the bird might be dead in a short time. Rather Leibniz was absolutely right to describe time as 'a succession', the succession of units of time. But events, as the real stuff of history, do not consist of units--- they are always in an endless chain. Given that we have to have a leader or ruler, we can trace the events from the beginning of life to the present about all our rulers or leaders as the story of our continuing live, and through that to all historical events.

To have time proper (such as can be mechanized in a clock), sentience is necessary: somebody must be there

What is time?

who knows the intellectual use of points, to count the orbits of the sun as years, and further pare a year down to our SI of time, or there will be no years and no seconds as fractions of the year. As any good book on astronomy can reveal, events in the cosmos happen haphazardly without order or logical direction. Of course there is causality, but no order such as time imposes on man.[49]

Thus, the logic of time in the universe will apply to any 'Beings' in any part of the cosmos. The parameters are (1) Regular or repetitive motions;(2) Chemistry and other events giving rise to a period of waiting;(3) A theory of numbers and arithmetic plus the ability to count---so

---

[49] A very important point about the 'Order of Time' which deserves a chapter to itself is this (and I hope it is noted well): The order of time seems very oppressive and unavoidable; indeed it is oppressive or unavoidable, because it comes from the strict and precise numerical units into which the earth's repetitive motions are (unstoppably) divided, so that every unit of time has its limited duration or range that it cannot logically breach---absolutely and impossibly. The seconds move on to the minutes, then to the hours, days, months and so on. There is not a single moment out of place anywhere. This is a very technical issue in the mechanics or mathematics of getting time into the clock; for time in the clock had a long and complicated journey from somewhere to get there. It is also one answer to Russell's query about what is measured by the clock: it is the space traversed by the earth divided strictly into units of duration so that we can live in accordance with the earth's conditions in safety. Everything people want to know about secular time is covered in these notes. Because the year is determinate and yet repetitive, each unit of time derived from the year has to have strict limit in range---that is the reason time is so oppressive. Each unit has a limited range it cannot breach.

sentience is required. I am excited because the clarification of time is going to open up so many aspects of logical thought for serious research, that's what I would call the triumph of science.

We must not speak of Man and the cosmos in any discourse----we are so infinitesimally insignificant and unsure of anything at all; the nearest equivalent in vanity and mystery is like a new-born baby telling the parents what to do---how did it get the knowledge or how did it know the language? It'd be a mystery. So is breathing mankind in the world of lifeless matter, except that we have the brain. With the brain we are able to probe nature all the way from the billions of galaxies to infinity. So what is the status of this frail and easily-destroyed man and his brain? The metaphysical status of our time in the universe is discussed in the book. The miracle is that we have these brains capable of probing the entire universe.

Another mystery is that everything begins with time; everything has to be in time or does not have the credentials of existence. Relativity, too, is relevant because it started the inquiries about the nature of time; before that we just took it for granted as fixed by God, it is absolute, generally covering the whole universe and the same everywhere, as Professor Sir Arthur Eddington put it (and will be quoted many times throughout this book because many scholars just cannot even begin to accept that that is the truth about time today.) Eddington wrote: "Prior to Einstein's researches no doubt was entertained that there existed a 'true even-flowing time' which was unique and universal...Those who still insist on the existence of a unique 'true time' generally rely on the possibility that the resources of experiment are not yet exhausted and that someday a discriminating test may be found. But the off-chance that a future generation may discover a significance in our utterances is scarcely an excuse for making meaningless noises" (The Mathematical Theory of Relativity, ch.1.1.)[50]

Of course, this is not a treatise on physics. The quantum and QED are mentioned only briefly here and there, though the thrust of the argument is directly related to the relativity notion of time introduced by Albert Einstein. And although this is not a book on physics as such, the idea that time is seen differently in science is shown to be completely mistaken. There is only one system of time in the world---it is the same one that scientists use in all their researches. Einstein changed time for all mankind not that he invented a new time system for use in physics or relativity, far from it. His suggestions led the philosophers to declare that time is secular and 'constructed' by man as will be explained in the book, and the quantum comes in because it is based on time; it materializes from a fraction of the second, and the second is also a fraction of the

---

[50] The italics are mine. For this book is based on his true, honest and categorical statement and Russell's judgment that time is a construction, and that universal time is abolished---which means we have to research how we obtained out time, and why the divine institutions would not even reply to my communications because I stress these truths beats my understanding.

earth-year, while the year is our basic unit of time out of which all other units are derived as fractions. So a new logic has been introduced and it makes time a) secular; b) constructed by man; and c) necessarily discrete.

All these dire inferences began (as will be made clear in the book) when the great Bertrand Russell deduced that according to the Einstein notion of time, time does not run through nature but is constructed by us and therefore, by implication, necessarily discrete. So cosmic time is abolished not only in science but for everybody else; it's been quietly ignored or neglected as an intellectual subject for debate because people----including the scientists themselves--- just cannot imagine that time is no longer fixed by god, no longer absolute, and does not run from the distant past to the present and the infinite future. That it's there even before we are born and leave it behind when we die, and so forth; all of which may be described as 'The Traditional Notion of Time'.

This is not a technical book on philosophy either.[51] It is rather something like "a general consensus treatise" about the ordinary thing we all know as 'time'---time for sports, for travel, for work and for doing everything we do.[52]

---

[51] One literary Agent praised my attempt to link physics to philosophy. That is his understanding. As far as I am concerned, I am merely making suggestions about the passage of time and how it is 'constructed'.

[52] Recently the Cambridge University Press (another divine institution) wrote to reject my book because it is addressed to the general public but their philosophy list does not serve that market. On the other hand,

What is time?

The sense of time, the entire human idea of time, is based on the passing days and nights. To know or understand time we learn that it is passing in days one after another, or nights; it is the same thing. This idea is then translated into physics, memory and the human imagination, metaphysics, religion, mythology, mathematics and ordinary linguistic practices with theories, customs and attitudes to match. No wonder time has become the most mysterious subject in the universe. We read time into everything. So when Einstein said it is limited to a frame nobody could understand him; when Russell also asked how we get it into the clock, he was ignored; and Whitehead's bemusing theory that it is instantaneous and therefore cannot move, was probably linked to Russell as 'philosophers preaching their usual incomprehensible abstraction unrelated to reality'.

In fact, they are all right---precisely as I have defined time above which, of course, means that what I write, to the vain literary gurus and academic morons, is not even worth looking at. Nobody has ever asked to see my manuscripts. They'd call for a summary and dismiss the work out of hand---some would even go as far as demeaning themselves by making rude remarks; and another begged me to go to the academic agents because not being a mathematician he could not understand a word

---

Cambridge have published (or foisted on mankind for its sins) a hefty tome about the arrow of time which does not even acknowledge that these assertions by Eddington and Russell, Einstein and Whitehead do exist.

I was saying! Yet I write in the plainest possible English language because I want to be understood.

The truth is that in the popular imagination and folklore, yesterday is the past, today is the present day, and tomorrow is the future. That is what all the religions and metaphysics and traditions are based on. From that we got the idea that the days and time as a whole are passing by. Yet there is only one day in astronomy! I am not the one to blame. There simply is no universal time, as Russell pointed out. I have thought about time without the obvious mysteries, because I think they are all false once time is seen as originating from, and limited to, this planet. That is what this book is all about; but there are defects in the book just because nobody would give me an ounce of literary assistance. It may well be they do not understand the mathematics by which it is presented!

Yet this book is not only aimed at the lay reader. The word 'lay' would make it an intelligent book for intellectuals who merely happen to be non-professional thinkers, the kind of people commenting on every subject in the letters columns of the serious newspapers. Alas, they are not the only readers. Everybody reads something and is therefore a book for everybody, since it is about time and time is exactly like life----we all have it, know it and use it. Nobody can do without time.

Before the creation of the Lorentz-Einstein notion of time, everybody relied on Newtonian time---general, absolute (meaning it does not change, fixed by divine power), covering the whole universe and the same everywhere. The non-religious general idea was that it just is, and even scientist thought that was a very clever way of defining time in logic! The new theory is that it is none of these eternal things as explained in the definition above---- there are as many times as there are planets, thus making everything that is said about time in the religions totally untrue. And the new theory is scientific because it is derived from objective experiments.

The reaction of the world as a whole has been to ignore completely the new theory of time that changes everything in metaphysics as if it has never been put forward, even though it is championed by Einstein, Russell and Professor Whitehead---the giants of twentieth century thought. And Professor Eddington also had harsh words for those who think otherwise. All this has been discussed in the book for that is where I come in, also to be ignored by the world as if I had never existed---except the abuses on the internet which I did not enjoy because this is a very serious subject.

One condition was that the new theory has to be in accord with astronomy, and I have shown that it is, because there is only one day in astronomy, an irrefutable fact in nature upon which my theory of time is based. I would even go as far as to assert that it is an undeniable fact without which any theory of time is flawed. The sun shines on us as "the day". Only one day constantly on; the sun does not go to sleep as we do during the night time, which is caused by the earth's revolutions. I am repeating these ideas deliberately and also charitably for the benefit of the largely religious academics in the universities. It is a pity the so called great academic institutions of this world cannot understand simple ideas about time alone, and continue to produce mighty tomes about time rather as if they're illiterate, praising Einstein for his theory of gravity, when, in fact, his theory of time was his greatest intellectual achievement, even Bertrand Russell---the modern world's Aristotle---thought so.

The Conclusion of this book is also long, perhaps too long; but I take the view that time is the scariest thing in the cosmos therefore any means to attempt its clarification must be given a reasonable chance of public debate. There are also four Appendixes (as thought expansions) culled from previous publications that I have used and re-used in many other books. I am very old (born 1938) and very ill and frail and get no assistance from any quarters.[53] I have only my son, The Crown Prince of Avatime, to look after me as my literary agent, editor and publisher, but he is only an engineer, although a saintly genius, with about 30 books to his credit. However I cannot worry him too much as he already works too hard. When you see a full-page feature article in The Guardian about any writer, you tend to believe that the Nobel Committee's urgent call is not far away![54] Thus I re-use many of my old essays in subsequent

---

[53] I have discovered that writing in old age is more strenuous than boxing!

[54] See "BUSINESS SENSE ON DEMAND", Guardian, 23rd Feb. 2007, p.5.

What is time?

books when I am too ill to write a whole one. A writer does not copy philosophy from anywhere but work it in his own mind tentatively until he gets it right if he lives long enough, and is able to work till his dying days. Thus in my previous ten books charitable readers with some knowledge of philosophy could sense that I was desperately struggling to say something about time that I believed was important; yet it's wrong to use that as the excuse for all those abuses on the internet, for after all I did not give up to go and write money-making stories about corrupt African politicians---of whom I knew a few---starting with the incorruptible J.J Rawlings, probably a descendant of James Clerk Maxwell who was cleverer than Isaac Newton! Rawlings is the Scot who set out to cleanse Ghana's politics and succeeded brilliantly; then he handed over peacefully and retired. When he left the corruption not only returned but reached the sky and stayed there. With a story like this if I could secure another brilliant Scot as collaborator, I would not need to annoy the wealthy aristocrats in CUP with my 'popular philosophy'! Another Publisher always replied to me saying 'Not suitable for publication by Penguin'. Yet they published A World without Time by Professor Yourgrau, in which book time travel becomes 'a scientific possibility'.[55] About time alone man is basically stupid. Its mystery and mysticism intrigue him most; thus

---

[55] I am afraid the theory of time, as Professor Eddington has hinted, is so mess-up that even the logic of it is mired in controversy, not that the writer wants it, but one has to attack and defend instead of just writing his prose.

religion will never fade since all religion is based on the time system in use. Yet for thousands of years it has failed to tell us anything intelligent, sensible and reasonable about the true nature of it except that 'Creation' occurred at a certain specific date---by whose time?

Nevertheless, seriously, I accept that (being an "uneducated African brute") I may have committed numerous literary crimes in my books and for that I apologies to readers; although if I were under an obligation to provide a defense, I'd plead that the only literary crime I know to have committed deliberately is the habit of using published materials to make up a new work just to bring out a theory that could not otherwise get published. In this book, for instance, the Preface/Definition, the Introduction and Chapter One (as badly designed as they are) contain many new ideas to justify this publication as a separate book. There is confusion everywhere about time and every work about it will no doubt reflect that. Nobody knows what it is, with some writers even claiming that it does not exist while using it daily.

The reason time is so exciting is that life and time are bound up together into one, whatever we do takes time, and life remains mysterious and inexplicable; people believe time, too, is just as inexplicable and mysterious---asserting, without evidence, that time has infinite past and infinite future, and they have all built their religions on this idea of time.[56] On the other hand, this book and the

What is time?

contents of my other books on time accept the scientific evidence (which, according to Professor Eddington, was revealed uniquely by Einstein's researches), that time does not flow through nature, is not the same everywhere and most definitely not eternal.

Russell also said we construct our time. So it means the days do not exist naturally in nature; we look outside and find that, in fact, the sun is on constantly, producing just one constant day. Man creates the days and weeks and months and years all the way to the millennia out of our own artificial and 'probably' false concepts. They are all

---

[56] In fact, infinite time past and future are plainly false---there is no infinite past culturally, and infinite future cannot be known to exist. History is the march of events not time, and the events in our cultural history are well-known. Even the scientific 'intervals between points' is not time because they occur to billions and billions of objects in existence. The same cannot be the single time system ticking away in the clock. But the concept of 'intervals between points', gave us the idea of inventing time for the clock based on the long intervals between the years, pared down to the seconds so as to cover everybody living within the year. All the mystery of time centres on the clock yet it can be explained in logic or with it, for the clock is based on astronomical events; every second is equivalent to a certain amount of space round the sun, the reason so many seconds equal exactly to one complete orbit of the sun---so that we can start another orbit the next second. It should be noted that the process is purely logical and yet amounts to something like physical oppression. For that is how things work: by means of logical deductions. The world is oppressively regulated not by physical forces but by logical sequences, even the physical forces have to obey logical sequences. Suspiciously, it looks as if the all-powerful human brain is similarly organized, but this not the appropriate place to go into that.

temporary blips of some objects over larger or smaller objects, plants and animals. There are numerous such blips all over; stationary plants, smaller animals and insects get most of them. For example, each time we walk we may block the sun for several insects---you cannot use that to interpret the world. The earth's blips across the face of the sun, which we call days and nights and multiply them to all the years and centuries, are the same: artificial, temporary and carry no metaphysical significance. To me this is exciting stuff, for proving that time is secular. It tells me a lot about the nature of life. No wonder like-minded Professor Eddington castigated all those who thought otherwise. I agree that it gives us a new meaning to human life on earth whether we like it or not for religious reasons.

Incidentally, as we try to strip time of some of the mysticism in which it's wrapped up, it becomes obvious that it's not true that whatever we do 'takes time' and therefore time (as the enabling force probably of cosmic origin), existed before we could come to be to take any such actions at all 'by time'. It is true that any action will result in the expenditure of time; it will also cause a period of waiting or connection to somebody (important in natural and laboratory chemistry), but not time for the clock ticking away perpetually. Time that we get continuously from the clock known as 'the time' is altogether different; it is not necessary before events; we've lived like wild beasts before without having 'the time' in a clock on the mantel, or the Rolex watch on the rich man's wrist. Rather time in that metaphysical sense is a concept (mechanized or not) that

What is time?

we retrospectively apply to events after they have occurred or even before they happen through planning. What is needed before any action is energy---whatever we do requires energy not time. But of course in special events we do use time to plan them---yet that has nothing to do with the explanation for the being of time in metaphysics. The myths of time are numerous; some of them are explained in the Conclusion below, but nobody can clarify them all due to passionate if not violent religious reasons.

To expand the point above about the essence of time in activities, it's obvious that everybody on this planet at some stage has been deceived by the many false theories of time, mostly from the religions, that make time so mysterious. I have myself thought in the past that time is inseparable from life, but now think otherwise. How strange it is that the mind can refine ideas after due concentration on a subject. After hearing from Russell that time is constructed and thinking about it deeply for several years, I have come to accept that time, once created and safely planted in the human psyche, becomes oppressive such that we can't do anything without it because it is linked to the earth's physical conditions to breach which is to curt danger, and physically fragile man has always been afraid of danger. After a while the idea of time has become instinctively oppressive so as to dominate life completely. All instincts and habits begin in the same manner.

Still time appears to be inseparable from life but not biologically or metaphysically, only culturally. In logical thought time is a concept formed from the conditions of the planet and therefore most suitable for the regulation of life as a whole in safety---even after death. Thus all existence consists solely of life, events and time. Life comes first, action follows and time is what we need unavoidably to regulate all activities, either before as in planning, or after in retrospect. Of these, only life is metaphysically important, the others are used to service or regulate life and human in origin---created out of need. As such, it is not correct to assert that "whatever we do requires time": the time is used to analyze events retrospectively unless we are planning them. We can live without time and we have done so in the past. Action does not require time but energy. So time is not so vital to life except that it can become a catalyst in natural chemistry as a period of waiting. But there is not much man can do about that.

The mistake of elevating time is cultural (or religious); and it has grown in the mind to become an instinct, or the habit of contrasting existence with time as a separate entity closely associated with all actions. Yet in logic time is a concept always added to events (or used to analyze them) only in retrospect or in advance. As a concept it cannot pass by physically but can progress arithmetically (like counting the years all the way to centuries, when, in fact, there is only one year repeated over and over again.) Since we count cyclical units as the rate of passing time (like the years) our time units are strictly determinate; they cannot pass by physically, I repeat; and in the absence of passing days as time all of the religious ideas about time become untenable in logic.

Indeed religion has never been a cause for good for mankind, just the glory of God. The benefits wrongly attributed to religion belonged to defunct and unrecognized thinkers, but even then their good works have been tarnished with the religious canker---and it's all because man is simply afraid of death. For looking back to what was annoying Professor Eddington in his day, and what has been happening to me now, it's beginning to dawn on me that my ideas and books are not welcome because nobody feels able to tackle post-relativity secular time as defined by Einstein (that there is no longer a universal time and every inertial body has to have its own time), or Bertrand Russell (that time is a human construction), or Professor Whitehead (that time is a sequence of non-interacting moments). There are a number of books about time but they all regard it as a mysterious entity that just happen to be there as "just is". This enables the writers of such books (published with great fanfare by the 'great' publishers) to dabble in subjects like End of Time, Time Travel and the Day of Judgment, etc. There is even a book called "A World Without Time" and it has been so successful that I was forced to read it. It concludes that under relativity time is impossible! Yet there is time. It is called 'Post-relativity Time' but it does not support time travel so that people could meet their grandparents even before they're married---such lunatic ideas. It also does not run all through the cosmos such that a second here is a second everywhere else, because the time is based on the repetitive motions of celestial bodies so variant that the units of time are bound to be different.

This time is of course completely secular and will die with the planet; for the planet is due to disappear; some people, (the majority of mankind) are afraid of that and therefore reject relativity time; I don't blame them. Time is closely associated with life, and scientific time is dry; on the other hand, religious interpretations of time are friendly and nourish human desires. The simple fact is that life should never have happened. It is a cruel irony that so sensitive a creature as man should come to life in such a harsh and unforgiving world, suffer all the pains of life and die in the end---sometimes painfully---for no purpose whatsoever. That is the reason post-relativity, dry and scientific time is unpopular. As bad as they are, the bitterness of human life makes the religions supreme; they struggle to make life worth living even with all their faults and myths, otherwise it's not worth it---although the Jesus Christ story is plainly rather shallow.

It goes against what we know about time. Of course no one can claim to know what it is, but we do know quite a lot about it. Starting with Einstein's researches, as Sir Arthur Eddington has confirmed, time is no longer seen as running all through the universe such that, in the words of Bertrand Russell, it can be applied "without ambiguity to any part of the universe". The idea (to emphasize it over and over again) is that time does not 'flow' through the universe; it is a sequence of non-interacting moments in the mind as waiting periods, leading to the instincts for periodicities.[57] Each moment is to be regarded as 'human

What is time?

contact'; how long it takes is the time. And we know that to be part of the brain mechanism human contacts could start from the womb. This may be speculation, yet it is based on an infinite wealth of deep knowledge since the days of Einstein. The all-important question is whether it can be used to interpret time metaphysically, meaning 'to its ultimate', and my answer to that question is yes. That is what this book is all about---namely secular time, 'constructed' by man, which originated from Einstein and Bertrand Russell. But time is so mysterious that people shake their heads when they hear this. Let me put it in plain language to help them: as a period of waiting logically time has to involve the brain (or a human brain), and material objects from which the relationship (the periods of waiting) can occur. And once this is realized the endless streams of mythologies about time come to an end---including that of Jesus Christ, His father and the day of judgment---alas there are even no days in nature at all but any planet can have its

---

[57] To me this sounds very convincing and the words for expressing the idea too are chosen pretty guardedly. As mysterious as it appears to us, logical analysis of time shows that it is almost exactly like money, always remembering that logic is the ultimate arbiter of truth: it is essential; we can't do without it; it is artificial; it evolved; it can be changed; it is based on known parameters. In the case of time the factors are space and the instinct for periodicities in the mind that probably formed in the womb----we call it the sense of duration, which means it is influenced by points, and the Whitehead notion of 'contact' or 'moment' is relevant. In the case of money it is influenced by human greed, selfishness and economic theory---the sharing of scarce resources, store of wealth, medium of exchange etc.

own number of 'days'. Yet God does not live on this planet so our ephemeral days do not apply to Him!

Actually, it is a surprise it took so long to realize this; it seems so elementary, so plainly doubtful that a universe of so many billions of bodies could be regulated with just one system of time, sleeping and waking all at the same times! What about the dark matter galaxies---how do they read God's time written with our bright photons? In logical thought or science the units of time are directly traceable as fractions of the earth's motions; and these motions, too, are repetitive, showing that the units of time are exclusively those of the determinate motions of the earth where man has overall intellectual hegemony. For instance, it has been known for centuries that roughly 31, 536,000 seconds equal one complete year. Not a second more---because we start another year instantly, whilst all other units of time are also mere multiplications of the second as our SI of time. Once it was shown that time is not fixed nor universally permeating the whole cosmos, the logical inferences that it is therefore secular could not be proscribed. Not any more by the religious leaders.

Nevertheless, Post-relativity time is only important for showing that it is neither fixed nor general---that it is not absolute for short. It does not mean that relativity has answered all the questions about time. That is not how our knowledge of phenomena occurs; rather it happens bit by bit---always tentatively. So the passage of time and many other problems have not yet been explained satisfactorily in relativity. In fact it seems nobody as yet has the answers.

What is time?

Nothing wrong with that; it cannot undo what has already been achieved. Einstein also failed (strangely) to include time as a postulate in special relativity. Had he done so we might have been mercifully spared the Minkowski fiction. For time is not just mathematical. It is evidently more complex than all science, astronomy and cosmology, language, logic, philosophy and mathematics put together, after all we can only study these subjects by means of man's use of time. The reason it is so painful to write about. Nobody believes that any one writer can know more about time than himself or the religions. Time is regarded as just is, or is how it is used. It is the only intellectual discipline that has no subject-matter but just how it is used. Strictly speaking, even that is unrecognized, nor honestly admitted. In logic and commonsense, language and meaning, the word 'time' is meant to be its manifestation in usage (known through usage), unless we bring in astronomy to show, for instance, that a second is 'so-many-miles' round the sun, but how many of us know astronomy or incorporate it in ordinary speech? That is not a reasonable expectation of humankind. Yet, without that, we use the word time senselessly---confused? Let me explain or try to explain. The rational basis of all human thinking is the Subject/Predicate system: there must be a subject, and what you say about it. You cannot speak about nothing; yet even if you try, you'd be speaking about something, namely 'nothingness'. This elementary version of the doctrine will do for now. But that is for thought, or reason only; for action, description or identification, there is always an object and what it is used for or its application (see my little book Time and the Application of Time). Time alone has only its application until astronomy is brought in. Then we can say Time is the accumulated number of seconds en route to a full orbit of the sun so that we can start another year (this is said normally in so many different ways, of course). For the second is our SI of time and all units of time are reckoned up or down in seconds. Even in the Night and Day aspect we reckon time in seconds or multiples of seconds to minutes and hours. (Hence I argue that the night and day system do not count in the metaphysical interpretation of time. It is also a fraction of the yearly cycle---the basic 'time unit' out of which all others are derived.) In this sense, every second refers to a specific position in the orbit of the sun and therefore substantial, covering space. For logically the clock on which we rely so much is not time per se. It is a device for telling the time only. We still have to define time as above or there is no meaning for the word in logic except its usage. However, since we do not speak as detailed above, time is always meant to refer to its usage to make sense. These are some of the technicalities in the interpretation of time that have taken me half a century to figure out and which those abusive lunatics on the internet social media do not know.

Thus, ignorantly, we call something 'time' and yet can only describe it from usage. The great and mighty in academia also forget that merging time with its usage in this stealth, instinctive or primitive way means linking it to space in mathematical physics again to constitute one entity (except by the 3+1 formula) is logically impossible.

What is time?

Since time has no subject-matter, how can it be merged with space when it is after all obtained as 'relations between points'? For once points are mentioned space is involved in the creation or construction of the time. Time appears weird, yet in logical analysis it is simply "the mathematical quantification of reality".

Those of us who insist on carrying on the thankless task of trying to explain time do so because the religious interpretations seem grossly unsatisfactory. One system of time cannot be applied to all sectors of this universe of so many billions of bodies, and in trying to trace how we 'construct' time, the yearly cycle came in handy, always remembering that logic is the ultimate arbiter of truth; and logically only the yearly cycle can act as our basic unit of time from which all other units are obtained as fractions. The important point is that by this supposition most of the quandaries of time (in theory, perhaps all of them) lend themselves to logical solutions. But again and again I must apologies that I am unable to present my ideas in traditional, 'learned' academic prose. So they say I am illiterate because the indigestible academic jargons are missing from my work! The fact that most academic books have benefited from sanitation in the hands of numerous scholars is no excuse; what may be regarded as good excuse is that time is difficult and so mythologized with thousands of fables, myths, legends and religion that it is approaching the point where it would be virtually impossible to discuss it in logic to aid science which needs it to be logically explained to accord with physical reality.

What I hope to have made clear in the interest of scientific thought is this definition repeated for emphasis because time appears to be life itself, with fiercely contested rival interpretations: time is the same thing as existence; it is "existence" **reduced to** small mathematical parcels for cultural convenience---culturally manageable quantities of existence created with mathematics and cyclical motions, like the year. So the year is one unit of time, our basic unit, pared down to the second as our SI of time. In consequence, all other units of time are multiples of the second. Hence time can now be linked to reality, for it is existence in essence except that it is parts of the existence created with mathematics for application in all events and it accords with reality. **So every unit of time is a piece of reality, a piece of our metaphysical 'Being'.** This is the reason time is virtually inseparable from life: it is an aspect, an intimate aspect, of life, small parcels of life created for our convenience in living, making it possible for us to attune properly with what constitutes reality 'out there'. If you are told to wait for ten minutes it means ten minutes of your life out of the total you could live is being expended---again, time and life are inseparable but the time consists of short burst of life created mathematically for cultural use. The mechanics and mathematics and philosophy are complex and probable untraceable, but they're secular because with the discovery of local time cosmic time was abolished; so 'there is no longer a universal time'. Of course there are factors, parameters and conditions in nature for the creation of time as 'a period of waiting' or in mathematical sequences in the clock, but it takes a human mind to do the creation.

The whole purpose of this book is to suggest that time is basic to everything----all existence----therefore we have to ignore the pretensions of mathematicians about cosmology, meaning all existence. I am not condemning everything they do; I agree that most of them are excellent mathematicians or scientific thinkers. But in the absence of cosmic time, we need to derive our mundane time logically from the reality spread before our eyes to give us the irrefutable theory we need to account for that reality, before we can be sure that it is really the _real_ thing, otherwise theoretical physics will always contain an element of mysticism about time and, therefore, everything else. This is, to me, a very serious matter ignored by almost everybody.

I repeat, reality in this world consists of the mechanics of bulky matter plus the mechanics of the quantum theory or QED---from the photons to the stars, etc. Yet the problem is that all of this is controlled by, dependent upon, predicated on or derived from time. So we must first establish what this time is, otherwise our theories in cosmology are mere speculations, especially since the Minkowski theory of space-time is sheer nonsense. Of course time is space-time because it is strictly dependent on space or derived from space as argued in this book, but that does not make it identical with space. You cannot use mathematics alone to equate space to time or time to space since we know time only in units. Points are involved in the creation of time in units---another 'proof' that it is man-made--- and where points are involved space is implied, making the equation of space to time absolutely impossible. Pure mathematics devoid of philosophy is barren. It is the reason Bertrand Russell was incomparable as a thinker no matter what his enemies are telling us.

Chapter One: The Scientific Notion of Time

The scientific study of time begins from the consideration of existence to reality, imagery, visual perception, events, regular motions that are reliably and metaphysically unstoppable, plus arithmetic, a theory of numbers and the ability to count, analyzed as follows:-

1. *Existence is the same thing as 'Being'[58] or 'being in the world as a sentient Human Being'; from this we*

---

[58] Calling the passage of time as the irreversible passage of existence has no meaning, because (apart from growth, which is chemical), existence does not move---yet we can speed time up without it taking existence

get or deduce what we call 'a conscious person' who has need of time and for whom logicians and mathematicians have 'constructed' time out of the elements of nature, otherwise "There is no longer a universal time..." This, in a nutshell, is the raison d'être of this book i.e. we are alive or living, and we create repetitive cycles and call them 'years', and further pare one year down to all the other units of time---all the way to the second as our SI of time. Thus, scientifically defined, time is existence reduced to shorter periods with mathematics and cyclical motions for our cultural, educational and intellectual convenience. Great intellectual efforts have gone into finding this out.[59]

2. *Reality is everything that a sentient Being is conscious of as existing or occurring in the world around him or her---even the word 'existence' may be used to cover all of that.*

3. *Imagery is visual perception and all the data of sight, obtained from contact with the particles of light, and not something supposed to come from God.*

---

with it. In any case, the passage of time has now been logically explained in this book.

[59] All the books I have been writing were attempts (perhaps imperfect attempts) to formulate a cogent theory of secular time, and I believe I've now got it right.

4. *Events refer to everything that happens to which repetitive cycles can be applied to determine their duration---that they were there for so many cycles, years or months, all the way down (in units) to the SI of time, and all the way up (also in units), to one complete year as the basis of our time reckoning. Thus all time is known in units as fractions of the determinate earth-year derived from orbiting the sun. Since the earth-year is determinate, of course, all units of time derived from it can only be in units. Also, since the orbits of the sun are unstoppable, time will always be unstoppable.*

5. *Repetitive or regular motions are the metaphysical cycles we count as the units of passing time. There is only one of these to use for time, namely, the yearly cycle upon which all our time is based and every unit of time is a fraction thereof, as well as being our measure of age and ageing---even though the true nature of time is unknown. Thus, ironically, the passage of time is what we call 'time': we know time from how it is used, and, because time is unstoppable, how it is used is how it is passing by.[60] This makes time the greatest mystery in life except*

---

[60] We cannot stop time in other to use it; it is passing as you are using it. We rather count it as it will pass forwards (or backwards), and use it as it is passing by. Again, we know time as it is used, and how it is used is how it is passing by. The reason is that it is the earth's motions we subdivided into our familiar units of time, and the earth does not stop for anybody.

the life itself. We don't get the year until it's passed, and it is the year we pare down for all other units of time. Above all it is indefinable. Nobody can ever know the temporal length of one year.

6. Arithmetic or the intellectual use of points, a theory of numbers and the ability to count the yearly cycles as 'years' and pare down one year to the SI of time (the second), and thus obtain all the units of time as multiplications of the second, so as to be able to divide existence into manageable units of itself (as it goes on, meaning as the earth goes on circling the sun), called 'units of time', the same thing as 'units of existence', for application to all events and images. That's how complex time is seen in scientific or rational thought, following the abolition of general, fixed or cosmic time by Albert Einstein. My message to all writers dealing with time in any form (including time travel and the multiple dimensions of reality even in physics), is that if they can't define time logically as sketched above, then they should be honest enough to admit failure. I read dozens of these and they make me sad. I used to reply to some of them, but stopped when I realized that they played mute because they didn't know what to say. There is no shame of failure in this kind of enterprise where we're all groping in the dark. I have been mocked for over fifty years, yet I'm still alive. Einstein did not delve into time. He took just one look at the Lorentz discovery of 'local time' and

called it 'time, pure and simple'. It was Bertrand Russell who interpreted that to mean cosmic time is abolished, and all time became secular as explained above.[61] To write about time as if it is still cosmic is nauseating. According to Professor Sir Arthur Eddington, time does not even flow, or is no longer seen as flowing through the universe, and yet they continue to write about time travel. For the last time, discrete time does not flow, it passes by through the replication of its units---and all time is known only in units, from the year down or up.

Let me stress most strongly, once more, that the most revolutionary idea about time in all history that has only come to light in the past hundred years or so (as stated in the notes), is that it does not flow through nature.[62] It

---

[61] Russell was the greatest philosopher since Descartes, and coined the phrase, "There is no longer a universal time...", to send the rest of us on the quest for the empirical origins of our time as attempted in this book, for instance. And since we have found the origin of time to be the point-divisibility of space, obviously, time is now seen as discrete and discrete time cannot move, run or flow---see below. The main problem of time now is how to persuade mankind that it is secular, discrete and therefore time travel with all the marvels they imagine is utterly impossible. There is no need for a PhD thesis about this. Time consist of units as limited intervals of reality. All our time units are fractions of the year, and the year is determinate therefore our time units are discrete. Discrete time cannot march, simple. If $s=ct$, then time travel could be contemplated (how wisely I do not know); but s (space) cannot be equated to time (ct). The Minkowski mathematics is flawed. He used two types of time in one equation, as $\sqrt{-1}.ct$. Yet time is one not two, as I have explained in the book.

What is time?

appears to flow (which will be explained in a moment), but if it can start from anywhere, then of course it cannot be running through in the form of a continuous thread, or 'flow' like a stream. The numerous suppositions in physics and mathematics about time are not for me to judge. I can only assume that most of them are accurate. But looking at some of the suppositions, I suspect time is still regarded as flowing through the cosmos and of which we tap our version with our mathematics; and if that is so then I would argue that either all of the ideas in physics and mathematics about time or most of them are flawed, plain fantasy, or irretrievably false, because I subscribe to the new revolutionary theory that time does not flow. So let me explain myself in plain language---all the logic and my reasons are implied above, yet I will state them again as clearly as possible though it will take some time, energy and space---quite unnecessarily.

As I have said, I cannot refute physics; nobody can do that except a lunatic. In fact, after years of cryptic silence, Bertrand Russell finally had to explain why he threw Wittgenstein out and never mentioned him as a philosopher or a serious thinker. He explained that his theories were all logical mysticism with the aim of putting an end to physics.

However we can criticize physicists or point to gaps in their ideas, a rich field for quarry because the subject is so

---

[62] So even the problem of the passage of 'flowing time' has disappeared.

vast and complex. One of this is the ubiquitous notion of treating time as 'just is', or that it just happens to be there, which means it is regarded in science in precisely the form preached by the religions. This is the reason Professor Eddington angrily castigated all those expressing doubts about the Einstein theory of time, which has been interpreted as 'a construction' by Russell and Professor Whitehead. It's all rather simple. Lorentz found time to be changeable through experiments, and he was not only surprised but refused to believe that it was time since time was supposed to be fixed and generally covering the cosmos in one format, otherwise no one could define it. So he suspected that there must be some mistakes somewhere in his research methods. However Einstein said there was no mistake and that time was changeable anyway. The first time in all history that anybody had theorized that time is changeable and not fixed by divine authority---I am sure the religious leaders realized what it meant for their philosophies. But I cannot tell if they're behind the neglect of secular time as an academic subject from which I have suffered so much! So many publishers do not even pretend that I am doing work of any value at all.

The next step was to try and discover how we get this time. This was where Russell deduced that time is a construction, created by man, and asked the question I have already mentioned, namely: "If cosmic time is abandoned, what really is measured by a clock?" Professor Whitehead answered him with the suggestion that time is a sequence of non-interacting moments or units. There is

What is time?

thus a great deal of logic, philosophy and metaphysics, including mathematics, in understanding what a second means in temporal terms.

All the time everybody including scientists (especially the mathematicians), continued to regard time as running or flowing through nature like some kind of a thread or stream---something which just happens to be in existence and passing through. And they continue to do so today. Yet by the new definition of time that is wrong. It is conceded that natural time or the sense of time exists as relation or interval between points---ultimately interpreted as contact between points giving rise to units of time, or the Whitehead moments. How we mechanize this to tick time away in a clock is the inconceivable conundrum in human history, yet it has happened. We have time in the clock, so the logicians set to work to analyze how we got this time system without divine authority, in other words if it can so ordinarily begin from anywhere.

In this respect, the most important aspect of time, it was found, is that it is known and used only in units. That being so, no amount of hard logical thinking could go beyond the idea that time units are gained as 'intervals between points', or 'gaps between points'. Russell put it as 'relation between points'. In other words, contact or contacts create units of time as successive gaps in contacts or perceptions---in the manner of the sounds of seconds ticking successively by a clock. The noise is not the second; the second is the gap between the ticking noises. Hence Professor Whitehead's 'moments'. Also the repetitive yearly cycle was a great help; so too were the features of the environment, astronomy and mathematics. I always put it like this: sentience, arithmetic, the ability to count and a theory of numbers, were absolutely essential.

Armed with these techniques, the clock was invented to tick units of time that incorporate knowledge of astronomy, environmental conditions, even metaphysics and mathematics. Every unit of time tells a philosophical story. That is the meaning of secular time as a concept we form out of contacts with objects for application to events, and so it is not physically in existence to physically pass by. Of course it does increase, but since it is known only in units, the units increase in procession to make it pass by--- like the years. Units of time increase in numbers to pass by and make time seem continuous. All advancements and growth in nature are caused by chemistry or accidents. Progress is not a natural movement because it is an ethical issue.

There is therefore no kind of insoluble problem about the passage of time. It is the first quandary in life solved with the secular time supposition, but there are others, instead of man's enslavement by the fixed and absolute time regime imposed by the fractious religions and their Days of Judgment----in fact, there are no days at all anywhere, let alone a special day of judgment. The whole cosmos is dominated by a single day which will disappear if the sun dies and we'd be covered by a single night. That is the physical reality in logic with which we have to live and interpret the cosmos, not just to satisfy our rulers in religion and politics. Man must learn to live by facts at all times not desire, religion or the imaginary notions that flash through the mind: to live by facts is wisdom; to live by desire is the path of doom.

In addition to all the attempts I have been making to make the definition of time without religion clearer (if the reader will forgive my sins), here is my explanation of how and why mankind got the definition of time so wrong as to say it is "The passage of existence". The fact is this: contact or intermittent contacts with objects is the sense we call time, especially the gaps between contacts; we quantify these with the units of time as explained in Appendix 1 below. As these contacts go on and things move, advance, grow, decay or even just sway, they are (I stress, still now), assumed to be the cause of time moving on and taking all existence with it. From the Cathedrals to the science laboratories, from Togo to Moscow, this is still the universal notion of time, without even the knowledge of what it really is.

The dissenting voices (Russell, Eddington, Einstein and Whitehead), are all dead and the problem is that all time is in units and in units only. This gave logicians the idea that it is derived somehow from determinate motions or cycles, and the regular years were cited, as the lawyers would say, as evidence. Of course, nobody understood this until recently when Einstein got the brainwave that time is limited to frames, or can begin from anywhere.[63] So eventually we have got the idea that to construct the clock to tract time for cultural use, the units of time had to be thought of, or actually created, as the fractions of the yearly cycle.

However it means the true nature of time is unknown. We count the years as units of passing time and pare a year down to the seconds, that's all we can do to track or reckon time; the cycles or repetitive motions are what we regard as the passage of time. Whatever real time is can never be known by man. Atomic time is not a new (or different) system of time because the oscillations are based on the second; so atomic time is also derived from the yearly cycle.

The crucial point is that the yearly cycle is physical not temporal, yet that is all we know as time, in fact it is our basic unit of time. Everything used for reckoning time is bound to be physical, so what is the time? To me, it is what we use to track time; we count the physical cycles as the passage of time. There is nothing else. Whatever we do time will remain unknown; even the duration of the year is not definable in logic without reference to other events. Nobody can tell how long the year is in temporal terms. So there are a lot of complications about time when we apply it to the universe. We thought it naturally existed and we just applied it; now we know that we merely count

---

[63] "All that was needed was the insight..." he said cheekily as if it was that simple.

What is time?

repetitive cycles and call them time units. And only time units are culturally useful; the fleeting intervals between points (though giving the sense of time), are culturally useless until they have been mechanized in a clock to tick away in specific units that accord with the orbit of the earth round the sun so that each unit will have a physical value and meaning in time---which time means what is happening on the surface of the earth. Round and round we state these things to give us time to live by without knowing what it is, leaving the religions and mystics to concoct their own teleological cosmic theories.

I am repeating these points on purpose because there are thousands of delusions about time so deeply ingrained over centuries that we will die with them; no one can correct us, but for a theorist, he needs to get his points clearly understood. God knows that the traditionalist will still abuse him, but there is nothing one can do about that.

Still we know something as 'time', and want to know why it seems to be passing, appears continuous and is linked to advancements. One answer is that it is based on the earth's orbits and they have never stopped; as a result the units of time are oppressively successive---that is the reason time waits for nobody. The implication is that the passage of time is real enough, except that the passage is not physical on earth but mathematical, as units of time in procession[64]---i.e. the year is only one, yet we have had all

---

[64] The physical elements in time are produced by the planet in the form

those centuries because the year is repeated over and over again perpetually---so also are the fractions derived from the year. Given the phenomenal speeds of atoms of light and so forth, images can be composed of units and still be seen as 'steady'. I have already explained elsewhere that the existence of photons negates the Platonic theory of Forms. Images can travel to the eye through the air unseen till they hit the eyes; they are no longer conceived as inherent in the mind to be activated by external causes. Thanks to the quantum theory, Plato's Idealism is not at all relevant. It does not even qualify as a worthy subject for discussion in philosophy only in mythology, mainly because the religions use it as evidence that God exist. Nothing wrong to say this in the churches to their flock and not use it to challenge and frustrate the rest of us in politics, social theory, education, science, philosophy and ethics, even the interpretation of history, the cosmos---and time!

There will doubtless be a number of subsequent refinements (hopefully in logic) of this explanation of time and the passage of time, but it is doubtful mankind will revert to the old idea of time as "the passage of existence", since this existence is multitudinous and do not all advance, move, grow, decay or die simultaneously or identically; some events in the quantum theory are even the reverse of

---

of the climatic events we incorporate into our time system---still as fractions of the yearly cycle---occurring at certain points in the orbit of the sun and converted to units of time: such as two months for winter, three months for summer, and so forth. Earth time is linked to astronomy for our own safety; time was never mystical, the religions made it so.

growth! Things persist, grow, move, advance or die through their own chemistry and accidents---see below for more arguments on this particular point.

Despite all this, it seems to me that presently publishers, scientists, philosophers and logicians all seem to be yearning for only dramatic theories of time.[65] But there does not appear to be any such magic coming from anywhere. Fantasy about time is hilarious because you can then twist time anyway you like, particularly time travel (or transmigration of souls) so that everybody can come back to life after death and death would not be the end of life.

There have been changes since Einstein as I will try to demonstrate in this book. However the irony is that time remains the same for everybody. **The logical or rational solution of the conundrum is to find an explanation that is logically consistent with time as we know it without any trace of mysticism and religion, which is now possible.** But so long as time from all angles remains the same (leaving the days and hours and minutes as they always are), the logical approach, being difficult, gets no hearing at all only mockery. I have already published a small book predicting that physics will one day have to change course due to the

---

[65] 'Time changes with speed', 'gravity slows time', 'time travel is a scientific possibility' 'you could meet your grannies before they're married' and so forth. Otherwise they don't want to know. Yet if cosmic time is abolished so that time becomes secular and therefore discrete, then how could it perform any of these miracles---without God?

theory of time---and that, I believe, will definitely come to pass.

Meanwhile, the Cover Story of the New Scientist issue of 15 June 2013, entitled "Space versus Time..."[66] caught my attention whilst in the course of writing my little book about Post Relativity time.[67] "And what is that monster?" you may ask, and you could be right. Many scholars avoid this Nobel Prize subject perhaps because it is difficult. I am not even sure I have done it justice.[68]

It is the view, originating from the Dutch Physicist H.A. Lorentz and adapted by Einstein as the basis of his theory of frames, I think, and amounting to the philosophy that time can begin from anywhere and therefore there are as many times as there are inertial bodies or frames. Strange as it

---

[66] Of late many academics are writing about serious philosophical quandaries as though they're minor issues in science to which anybody can make a contribution. Yet philosophy is unique and remains extremely important---not a field for small contributions but a rounded view of nature.

[67] People still believe that time is just is, not that there is now pre and post Einstein notions of time, yet Russell and Eddington stressed this all their lives. I am not surprised that I am ignored for even they got nowhere. Human beings are not good otherwise politics won't be so messy. The only man to know more about human nature was Freud. In an area of high intellects, high stakes and phenomenal rewards, nobody would admit that you have said anything memorable---they would rather kill you and steal your Nobel!

[68] This piece was originally written as a journal paper so many of the references overlap with other notes in the book.

may sound, that single scientific discovery forms the basis of the new definition of time which sees it as discrete and not, definitely not, running smoothly all through the universe from the distant past to the infinite future, since, logically, 'the distant past' is owed to memory as history, and 'the infinite future' is fantasy in astronomy.

Nevertheless universal, divine, absolute and fixed time for the universe are all abolished under relativity[69]---a very serious intellectual challenge requiring the redefinition of almost everything, including religion, the Day of Judgment and life itself because without time there can be no life, since every object and every life has to have its "when" of existence. The most logical definition of time is "From when to when." It implies existence. <u>Nothing can exist without its "when" (that is the moment or time) of its existence. And if this time is not naturally existing to which all mankind is born, as an objective entity, then serious philosophical issues arise</u>. But the passage of time is not the passage of that existence. Yet we have lived without time before. It is rather additional, secondary, or complementary to the existence. Something we invent for use just like machinery, not something we move with physically in tandem.[70] We remain where we are (you may stand or sit on one spot for hours); only the shadows moving over the existence

---

[69] That is the reason for calling Einstein a 'Philosopher-Scientist'.

[70] Time's equation with life is religious in origin and probably Pythagorean.

constitute "time"---and only the shadows move, not the existence. Growth is another matter; it is caused by chemistry. Above all, time is nothing like ordinary motion. We use repetitive motions merely to show how much time is passing even without knowing the true nature of time; for if ordinary motion constituted time we would have millions of time systems without a stable one to live by. But once a system of time is in use, every activity, motion or even just sitting still can be interpreted as expending so much time through the arithmetic replications of the units of time in use---seconds, minutes, hours and so forth.

Furthermore, how can anybody sit still as time passes by even though time does not move? The reason is that something is moving to give us time, but what is moving is not the time itself, being the successive hours or days. It is an instinctive mental illusion to think that time moves physically. It does advance but only arithmetically through succession. Time is a moment of contact with the external world (vision, tactile, etc.) If the contact or perception of the external world continues---say the length of daylight---the time arises from counting something to give you how many 'somethings' (or numbers of it) you counted during that perception or contact with reality. Each unit of time is on its own but in procession with other units, though unrelated ('non-interacting units of time' is the logical name given to this process by Professor Whitehead). What are moving are the shadows of the astronomical bodies, since they constitute the time system. Thus you could sit still for hours, meaning counting the cycles or the regular motions used to reckon time. The motions are physical not temporal----but in psychology we count them as the rate of the passage of time. This is probably wrong, but we've got used to calling it time, and is the nearest we can get to defining time in logic.

Contrary to the concept of time passing through as the irreversible passage of existence, actually the passage of time is the time, for the passage of existence does not happen. We're misled into thinking that the motions of shadows that constitute time means existence itself is being pushed by time, moving as time or moving with it. This cannot be correct because existence is not one. It is multitudinous and move erratically---some are moving up, some are moving down, in reverse or even flying, but time moves steadily as programmed (in seconds or hours). These physical movements of existence (of Beings and things) are also purely physio/chemical and can only advance through chemistry. For existence to move in the way that time moves, all existence would have to advance by the same means---i.e. either in seconds or hours, days or months, as the case may be, yet that clearly does not happen and is quite impossible. Multitudinous Beings move in multitudinous ways. For all objects to move in one prescribed manner they would have to be identically constituted in chemistry which is quite impossible.

Looking back, I also recall another article in the same Magazine by the great science writer, Dr John Gribbin, entitled "Pay Attention Albert Einstein!" (New Scientist, 2nd Jan. 1993, p28---see below). These two articles have put in print so many controversial and contradictory assertions about time or post relativity time that I am lost for words. Yet once something is published it's almost impossible to refute it, otherwise Idealism would not still have millions of followers in the churches.

Dr Gribbin asserts that the Minkowski formula helps the understanding of the special theory of relativity. He said: "Minkowski's geometrical description undoubtedly improved the clarity of the special theory and is still regarded as the best way to understand it."[71] He's not wrong; everybody thought so at the time, yet Minkowski came in years after the world acclaim of the special theory of relativity where space and time, in the words of Professor Bernstein, "were made separate and played distinct roles".[72] We are now told that Einstein did not even understand four-dimensional space or 4-D geometry, and I believe it---see below—for he had his own definition of space-time to begin with. There is no doubt that Einstein was very, very, very, clever! It's unlikely we'll ever see another like him. Not in the age of the Internet anyway, when the half-wits are claiming that robots are going to take over the control of human life, assuming (wrongly) that any robots could do that without microchips of them being planted in every man's belly. And maybe they could

---

[71] Ibid, p28.

[72] Albert Einstein and the Frontiers of Physics, Oxford and New York, 1996, p110.

What is time?

do that to their own bellies, but I'm going to insure mine against that.

With Albert Einstein, it's different. We're still learning to understand relativity properly as the essay "Commentary on The Theory of Relativity" in The World of Mathematics makes clear.[73] First and foremost, time is not running through nature, as Professor Sir Arthur Eddington has categorically confirmed. Also there is no longer a universal time (Russell)[74], and time doesn't seem to move at all. And if that is so then time is our own creation/construction, Russell again)[75], and must necessarily be discrete, for

---

[73] Published by Tempus Books of Microsoft Press, 1988, Vol. 2, p1083: In this essay, the author makes it plain that we do not as yet understand relativity properly, so much for the British halfwit cosmologists who claim to be as clever as Einstein by merely speculating about the consequences of black holes---and adding insult to injury by saying Minkowski helps us to understand relativity. Actually he distorts the theory. All those who propound theories based on 4-D geometry are distorting relativity because it does not exist simply because time cannot be represented with 'i' in mathematics, let alone reality. So let us try to forget about the Transformation of Coordinates as it could end in fantasy. The Minkowski ict equation cannot represent true time, because time is not imaginary. That's the end of the matter with Minkowski, Time is difficult to define but it is by no means imaginary. Besides when you find a cogent logical definition you realize that it is required in all human activities, from going to the farm, work, travelling, schools in sports and even sleep, scarcely something that can be considered imaginary.

[74] From his ABC of Relativity---see below.

[75] From Mysticism and Logic as cited below.

Lorentz has proved by experiments that it can begin from anywhere. Let us clear up some of these points at once.

ASTRONOMY AND TIME

The concept of time in experience may not be astronomical, yet it is caused by the motions of astronomical bodies. Take the sundials for example; they do not move but they can tell the time as moving unit by unit---they are units of our own making, which is another evidence that time is 'a construction'. Otherwise what is the time they are telling, or giving us? Where does it come from if there is no longer a universal time? This was the greatest question about time or post relativity time asked by Lord Bertrand Russell---the world's most recent greatest philosopher, as already mentioned, but worth repeating because intellectually it is the most important question about all existence, man and the whole cosmos included. He asked in his book ABC of Relativity Ch 4, (but of which most writers seem unaware), "If cosmic time is abandoned [just as it was then being announced] what then is measured by the clock?"[76] The logical or scientific answer is

---

[76] The original query is this: "If cosmic time is abandoned, what is really measured by a clock?"But I regard it as one of those clever sayings that can be stated in many variant forms with the meaning intact, and I enjoy teasing those half-wit scientists and mathematicians who fail to study the philosophy of science in depth. It's a serious issue, and no laughing matter. As a scientist you must start, as a universal duty, by asking why

nothing, absolutely nothing. Thus some writers claimed that time cannot exist under the special relativity metric---but we do have time! And we do live in a special relativity world or frame. Yet the irony is that we do have time under relativity.

Since this is the most difficult subject in our intellectual life, let me stress the bone of contention again and again. If it destroys the beauty of my prose, so be it; we need to be informed not entertained---the TV companies and films will do that well enough. Intellectually what we want to know is the acceptable logical definition of time or post relativity time, if time is not fixed, divine, absolute or generally covering and running through the universe and the same everywhere so that a second here is a second everywhere else. There! And we do know that this is what relativity time means---namely that it is not fixed or absolute, divine or general and the same everywhere. It varies from place to place according to the Lorentz discovery of local time, also known as t_1. I can now see why Einstein originally wanted to call Relativity 'The Theory of Invariance'.

As stated above in the notes, Einstein was not called Philosopher/Scientist for nothing. Indeed Bertrand Russell considered his theory of space-time as (perhaps) his most important revolutionary discovery---and God knows there

---

anything is what it is if not created by god. Finding the answer is your scientific research, though the question is philosophical; so anybody who disparages philosophy is lost. But to know philosophy you must start with logic.

were many others. To me it is more important than his explanation of the cause of gravity. I don't care much for interstellar gravity; the stars can go on swallowing each other; it has nothing to do with us, but time has. (It is worthy of note that his concept of Space-Time was different from the Minkowski theory---for space was merged with time in the special theory of relativity but by the 3+1 formula: that is, physically not mathematically on the basis of imaginary time.)

Astronomy is what concerns us, not cosmology. That is an interesting intellectual game mathematicians like to play (an indulgence) to entertain themselves when bored or lonely. Even if it can affect us there is nothing anybody can do about it. It's often forgotten how insignificant man is. In the wider cosmos we don't count at all. Even our sun, as huge as it seems to us, is really only a minute infinitesimal dot in the Milky Way and yet there are billions of such Milky Ways all over. This is the universe of complex events occurring in all sorts of bodies, some of them so huge that millions of our petty sun 'can find room in them'. As I write it is reporter in the papers that astronomers have discovered a black hole about 12 billion times more massive than our 'huge' sun.[77] So what is the status of this little animal called man? Philosophy is interesting and should be taught in High Schools.

---

[77] The Times of London, 26th Feb. 2015, p20.

WHY DURATION RESULTING FROM CONTACTS IS THE CAUSE OF THE SENSE OF TIME

Again, to avoid confusion and misunderstanding, I will state the secular theory of time upon which the book is based: it is agreed that the sense of time is a period of waiting, or a gap in sensation; it is known as the natural sense of time, or simply as 'Natural Time'. Any successive intervals at all constitute time or successive units of time. If they are long we divide them into separate sub-units, or of gaps also called 'time'. Yet everybody knows that that is not what we know and use as time; not the time we mechanize and distribute throughout a community; it is not, in short, the time for civilization that we can teach as part of astronomy, (orbit of the sun), and a logical succession of temporal units in sensation ("non-interacting moments")[78] usable for planning, retrospective applications, analysis or the scientific and philosophical interpretations of physical reality. This is a long definition of what we know and use as time. It is long and complex because time is the most complex phenomenon without which life and civilization would not be possible---and yet it is no longer seen as cosmic or divine but constructed by man as a secular entity limited to a frame or planet. What has been overlooked is

---

[78] A moment obviously incorporates consciousness and should be understood as any contact, reaction to stimuli and vision. Philosophers use the umbrella word 'Perception' of any kind. Thus 'a non-interacting moment' means a time unit of whatever duration.

that even in this secular and mundane sense it plays a vital role in our existence. So, ultimately, it's back to the accidental collocation of atoms, as Russell put it, responsible for all the creations of the human brain and of which time is probably the most important part. This is sad because it dissolves human vanity about our noble cosmic birth.

That said, all the elements that cobble together to give us the sense of time passing are connected; but this is just like saying all things in nature are connected, and also all things and events are caused. So back to the sundials; it was necessary to clear up some related points so that it can be clearly understood. The sundials tell the time by means of shadows. But that is all we can ever know of time and never the true nature of it; the time is driven by something, the real cause of time, but the nature of the ultimate cause can never be discover absolutely clearly either, because it consists of several factors, and at any one time we cannot tell what is doing what to give us the time.

The time a sundial is telling or showing is obtained from the duration of the earth's orbit of the sun, or the act of dividing the earth's orbit of the sun. 'Duration' means "during the time it is there" or through historical records (the same thing as 'experience'.) Let us use the simple word 'duration'. Since it means during the period of some event, it is always used with reference to something else or other events. The orbit of the sun is one such event in human experience logically suitable for marking time; and we divide it into units of time (by space traversed) for cultural

What is time?

use---from seconds to minutes, hours and days, etc., and the sundial shows how these units are passing, and we are forced to obey them because as they pass they bring darkness and the rest of it that could spell doom. Hence we read the passing shadows as units of time and act strictly according to them. They are, in fact, units of passing space, but we call them 'time' or units of time; and that is how we get time to live by. For the more we probe time the more it becomes mysterious; to make sense we have got to stop at a point that is logically definable and say we simply do not know what the rest is. That is the best we can do through logical thought or science by courtesy of relativity---though presently ignored due to the influence of man's religious pretensions.

Thus a sundial is a clock; it tells or reflects what the time is. Yet a clock which tells the time or shows what the time is solely by means of shadows means it is recording the passage of shadows as the passage of time. So the shadows are giving the time—and constitute what we call time in essence. Since there is nothing else to call time, the shadows are, in fact, the time, all that we know as 'time', and they are passing **so the passage of the shadows is time or the passage of what we know as time,** without any need for theories to account for how time passes through nature. This is my proof that all we can ever know of time is how it is passing by. But of course I argue that this passage is the mere replications of the units of time in procession and not as some kind of a thread running all through the universe. That idea cannot be proved, and what man cannot prove he attributes to God; on the other hand, we know time in units only, so we can deduce that these units in rapid procession can cause time to run through or pass by---and the speed of action on the atomic and quantum level is so phenomenal as to prove the theory correct.

An additional point is that the shadows are caused by something close by, the closest causal agent may be called the 'primary' cause. Let us call the cause 'duration' for that is what gives us the sense of 'a period of waiting' that can be divided into shorter and longer periods. For instance, daylight is caused by the period 'during which' one's part of the earth is moving across the sun; if the movement is slow the daylight would be longer than the normal twelve hours; if faster, the day would be shorter and so forth. The duration of the earth's journey across the sun causes the time.[79] Thus there is no argument against calling duration the power that causes what we experience as 'a period of waiting' or time. When we are lucky we can identify the primary duration (that is, the nearest.) Otherwise it may be hidden, or seen to be caused by obstruction, motion, inertia, ebb and flow, etc. ---and ignorance; for it is ignorance to say we have days of the week: there is only one day. The rest are human concepts for cultural convenience. I must mention here that cultural facts and

---

[79] As hinted above, duration, like the year, can never be defined except in reference to something else. Although unheralded duration is the greatest conundrum in the universe. To my mind it is existence but consists of layers upon layers to infinity.

What is time?

natural facts are two different things; one is for physics and philosophers, the other is for human convenience.[80]

Thus we can imagine that there are other causes behind the primary duration, and on and on to infinity (Primary, Secondary, etc.) They do not concern us; what matters is that the sundial is a clock which tells what the time is by means of shadows; so when we read the sundial as saying it is ten O'clock going to eleven O'clock it means the shadows are moving and that to us is 'the movement' of time. So the movement of the shadows used to tell the time is what we mistake to be the movement of time---and time itself, whatever it is, does not move. However, since we know from QED that everything is caused, we can assume that the shadows are driven by primary, secondary and other duration, also caused by other forces to infinity. If cosmologists want a logical method for defining time without mythologies, here is it! Duration comes in because it is what we can divide and mechanize so as to be able to tell that one hour is longer than one minute.

Let me repeat for emphasis that the shadows that sundials record as the passage of time do move, of course, but not the sundials. So it's astronomical motions and shadows that we use to reckon time. Time itself does not move, as it's a mere psychological concept---i.e. counting

---

[80] I agree that this is not a theory academics can respect because it is far ahead of anything they know; but at least they should reply to people's letters, since not all of us are humbugs.

the orbits of the sun as years is a matter for psychology not physics. We don't know what time is because the orbits of the sun are physical not temporal. We count the orbits to indicate the rate of the passage of time and never the real thing; so the passage of time is all we can ever know.[81] Logically it's impossible to discover how long the year is in duration. Nobody can define the year in logic without using any of its fractions; but then the fractions too have to be logically defined. It's therefore impossible to define time; and since the time units are created by ourselves out of our perceptions, Whitehead was right to consider time as 'instantaneous spread of the apparent world', <u>since the units of time we use or need are created with points out of the moments of time perceived, as already mentioned</u>.[82] So the best we can do by way of definition is to say time is the mechanics of shadows moving over existence, just as it happens over sundials to tell the time: you stay where you are and the shadows tell you that you've been on that spot for so-and-so cycles, hours or even days or years, like

---

[81] One orbit, for example, means a toddler should at least be able to stand on his own, even if unable to walk; from such instances, we work out what orbits of the sun mean to us in terms of time: one orbit is so-and-so, two is that, and all the rest of it down to the second---it does not give us anything else except that we call it time. In fact, it is the rate of the passage of time only. So there is no need for theories to account for the passage of time, as I keep repeating and have to repeat to dislodge that aspect of our minds conditioned by the primitive notions of time.

[82] A mathematical equation for this is provided below.

What is time?

houses remaining where they are for ages. Ancient man was wiser than we thought.

We don't know what time is (and it seems nobody can ever find out) but I suspect that, as a period of waiting, it has something to do with physical and organic chemistry, the natural processing in nature that causes us to wait (for some time in any activity.) The waiting period is what we call 'time'. For the waiting period is divided by the number of passing cycles to reduce it (however brief or long it may be) to units of time, and since these units are fractions of the year obtained with points, they can be called 'cycles', scarcely different from tapping the figure. But I hope the reader will also realize that I am repeating certain points for emphasis as the mind is otherwise so steeped in ancient myths about time that people would not understand what I am saying, set against what they call 'time', consisting of past, present and future to infinity, which is nonsense. After all even serious academics say Einstein was wrong about time because he dismissed past, present and future as 'stubborn illusions'. Man is basically silly because of his emotional problems; for most of the puzzles in life may be soluble, yet men are emotionally prevented even from considering some of the solutions because they do not like the look of them; and unfortunately that's when religion comes in. Whenever man is defeated by natural forces he calls on his God as 'The Father' for salvation; but once religion comes in death to the heretic is not far away, hence man's life is bedeviled by warfare.

Samuel K. K. Blankson

WHY SECULAR TIME IS NECESSARILY DISCRETE[83]

In sundials the motions of the shadows mislead us into thinking that it is the time that is moving. This is an excusable misconception because time goes from unit to unit, hour to hour, and year to year. Since the time units are fractions of the year and the year too is determinate, the time is bound to be discrete overall. The mathematicians were right all along, except that they didn't know how right they were and thought they're copying it from God---and all because of their feelings about the existence of God! That is the kind of human emotional burden I was referring to a moment ago. Many scholars are so consumed by their religious beliefs that even though they accept time as secular (after the Einsteinian revolution), but still regard time as running through the cosmos from the past to the present and infinite future (in phrases like 'the dawn of time' or 'end of time'); so people can have hope that time travel might be feasible. If they understand the Minkowski formula they additionally conclude that "curved space-time" makes time travel a certainty. Yet our discrete time is based on the yearly cycle,

---

[83] The mathematicians who created the SI of time linked time to astrophysics and the orbits of the sun empirically without knowing what they had achieved--- how, for instance, it rendered all religious ideas about Creation and purpose of life false and delusional---and we had to wait for the discovery local time to abolish cosmic time, even then not successful yet!.

What is time?

for discrete time means from one point to another, and we always have to repeat the year to make our time continuous.

Discrete time cannot move; discrete time cannot march forwards or backwards. It consists of moments of contact---long or short. Some of these are so long that we count cycles to know how long they last as they recur over and over again---like the Day and Night system. In short, Professor Whitehead got the philosophy right. From Lorentz to Einstein, Russell to Whitehead, and the quantum theory to QED we can be fairly certain that we know what time is but very difficult to demonstrate plainly and a lot of scientific imagination and deductions are required. Not just thinking but thinking scientifically or strictly logically despite the lure of mysticism where time is concerned.

Again, in reality, only the 'discrete' cycles we use to reckon time do move; time itself does not move, so it cannot travel all through the cosmos. We don't even know what it is. We count the cycles or regular motions as the rate of the passage of time. However those who argue that there is no time have a point, except that they do not make the point clear, or define it logically. For it appears that the movements we call "time's movements" are caused by the shadows of astronomical bodies, exactly as demonstrated by the sundials---that is why the passage of time is the passage of these shadows. Or the passage of the shadows is felt by us as the passage of time, since they bring the days continuously and the passage of the days has long been known as constituting the passage of time. In reality, there are no days in nature at all as an act of temporal advancement because theoretically we can have several days and nights in the period it takes to cook an egg. There is only one real day of significance in the reckoning of time for ageing purposes.[84]

The same shadows create the illusion of the days succeeding one another. And the hours of the day (or specific units of time) are caused by the slow movements of these shadows across the sky. All we do is subdivide them with points and give them names for cultural purposes--- from seconds to hours, and the days too from Sunday to the next Sunday, yet as observed above, what is cultural is not necessarily present as such in nature. What applies to the days, applies with equal validity to the years as part of the logic of time in the universe; every existing sentient 'Beings' will have to create a time system similar to the one described. Otherwise time does not exist; it is 'constructed' by sentient Beings. The factors and parameters required for constructing time sequences may be everywhere, but it takes intelligence to do the construction.

---

[84] For the whole importance of time is that we age 'over time' and believe, wrongly, that time is causing the ageing, pulling it along or something like that. But now we find that the time is not physical but merely a conceptual thing we construct to guide our activities, and that is what is proving difficult for mankind to swallow. Yet ageing is caused by chemistry, so that you can age a hundred by chemistry during the same period it takes to boil another egg!

What is time?

When writers talk of time and the cosmos, they are using earth time without knowing it. Since time cannot travel through the cosmos, they carry earth time in their heads which breaches the Einstein theory of frames. There is no time in the universe. Outside a human head, everything in nature depends on random chance and accidents. We make sense of events with the logic of time in our heads; so it means we influence nature, and further imply that complete objectivity is impossible. That is the truth, and how it affects us is unknown, or not yet determined. Again and again we are reminded of Plato's parable of the cave.

In my philosophy---obviously a fallible interpretation of time from one man---all time consists of units which replicate to pass by conceptually, not physically, and they are all fractions of the year, as our basic unit of time. Alternatively, time is seen as connected to consciousness and a portion of space and reality which every person inherits as part of his mental make-up for application to nature in order to be able to live; with mathematics we've cleverly created culturally manageable units, with the earth-year as the basic unit, and the second as our SI. In other words, we use mathematics to massage reality for the manageable units of reality as our time units to live by---that's another way to define secular time, always remembering that secular time is time that we can deduce from logic, as opposed to providential time, which is pure fiction. So all time is secular time. It's going to take hundreds of years to understand the true nature of this time, after I've been killed off with mockery and personal humiliation and insults. At least they have not asked for my head yet!

However that may be, there are only three units of time anyway: the day, the year and the second, meaning there is only one day, one year and one second. All other units of time (as fractions of the year) are multiples of the second; that is why one second before midnight is still today, and one second after midnight is tomorrow. The same thing applies to the year on 31st December: one second before midnight is last year or this year, a second after midnight is the New Year or next year. Thus the second is as important as the year. The daytime is the most important unit of time enabling us to live properly in the world, and at all times the sense of time or duration is a period of waiting as calculated in the mind of man---the universe itself does not act according to time or logic. Order and time are imposed by us when we deal with nature in whatever venue: "there is no longer a universal time...", according to Bertrand Russell, and I agree entirely. We all have to agree with him, because Lorentz has discovered local time, meaning everybody can create his own time, and therefore cosmic time is abolished. Let the religions preach all the sermons and incantations they can invent, time can never seem to be cosmic again, unless we want to reject all science and logical thought.

Also we normally think the greatest mystery of time is that it goes on and on. In fact it does not. We go on and on till death and we carry our notions of time with us.

What is time?

Metaphysically time is one vision, consciousness, contact or existence at a time, except that it is repeated---hour by hour, minute by minute, year by year. We can't have two hours at a time. Similarly we can't have two years. They are units of time repeated indefinitely so long as we continue to live. Death ends it all; so when we're all dead time will cease to exist. Nobody will be there to count the orbits of the sun as years and pare the year down with mathematics to the second as our SI of time.

Once the days are eliminated from the reckoning of time (since there is only one day), the weeks and months can be swept away; so we can all see that time is not marching on in days or months.[85] There will remain only the year and the second. The year is pared down to the SI of time as the second, and all other units of time are multiples of the second, and they replicate for time to continue or advance. Time does not pass by at all. Time units just go on replicating themselves. Age and ageing are chemical in nature and causes. Our traditional notion of time is

---

[85] Of course time does seem to advance, but since it consists of units, the obvious method is the replication of the units. We cannot see the process happening; all we see is the smooth advancement of time as if it is moving across space. But the logicians objected that, since the units of time are fractions of the year, and the year is determinate, time must be discrete---and discrete time cannot march through nature. For the Cambridge University Press to reject this reasoning because it is popular philosophy for the man on the street is a statement that the CUP is unable to understand lucid philosophy except the meaningless garbage written in obscure language like the stuff of Wittgenstein.

religious and completely wrong. The most logical definition of time is Whitehead's "sequence of non-interacting moments." These cannot march through nature. There is only one year and only one second, replicating on and on indefinitely----time does not move, march or physically pass by. It is a concept from human contacts linked to the motions of the earth round the sun and mathematically divided in such a clever way that a precise number is equal to one year and we start another year. In between, multiples of the SI are found convenient for cultural purposes. The idea that time passes together with experience or existence (as the irreversible passage of existence) is completely wrong. We and our creations (infrastructures, buildings and so forth) are never in motion. We always remain where we are; also age and ageing are chemical not temporal.

As I have said before, the best logical definition of time in science is "a period of waiting"[86]; but something causes the period of waiting otherwise there is no time as a separate entity. What causes 'the period of waiting' is time in the eyes of the person who experiences it. It is also time to the person who causes it, if it is caused by a sentient being; the need for periodicity is a natural law (and probably originates from the brain), and, to me, it is time; so it is not correct to assert that time does not exist---we

---

[86] I borrowed this phrase from Professor Richard Feynman.

What is time?

couldn't even live without it, nor could we define existence without it since to be is to be in time as explained below.

We count the mere shadows or cycles of physical movements of time for the real thing: we say we are aged ten years when we orbit the sun ten times---but the orbits are mere physical events! In the term 'year' man has equated linear space to time without knowing it, and we continue to suppose that the year is a cosmic creation. It is our most spectacular intellectual failing. By this new theory, indirectly, even the problem of the passage of time is solved also without knowing it. We think know how it is passing; but our knowledge of the passage consists of our counting of physical cycles, e.g. as 'years' pared down to the seconds and atomic units. So the passage is the time, if I have to repeat it.

Altogether, it's astonishing how Einstein was able to solve so many of our dire problems in so short a time. I believe it's because he's strictly a logical thinker (as Russell repeated many times in his many books), and through logic and the dynamic mechanics of atoms and mathematics, all things are connected; therefore science is indispensable in all human affairs, not just one way of looking at the world but the only reliable way in so far as all things consist of atoms. Even this last sentence is inferred from Einstein's ideas. So long as there is physical elements in anything, there is bound to be some aspects of science in everything.

As noted above, there is no longer a universal time covering the whole cosmos, which to me is the most serious philosophy of nature since Copernicus. This is how

Russell put the notion: "There is no longer a universal time which can be applied without ambiguity to any part of the universe; there are only the various 'proper' times of the various bodies in the universe". For Einstein, (as Abraham Pais put it in his book "Subtle is The Lord..."): "... there are as many times as there are inertial frames. That is the gist of the June paper's kinematic sections, which rank among the highest achievements of science..." He went on to plead that it should be taught in schools, and I agree with him.[87]

Next, Professor Sir Arthur Eddington. As quoted above already (probably angrily) he wrote[88]: "Prior to Einstein's researches no doubt was entertained that there existed a 'true even-flowing time' which was unique and universal...Those who still insist on the existence of a unique 'true time' generally rely on the possibility that the resources of experiment are not yet exhausted and that some day a discriminating test may be found. But the off-chance that a future generation may discover a significance in our utterances is scarcely an excuse for making meaningless noises".[89] Given these categorical notions of time as a secular entity, the definition of time has become very philosophical and difficult, making it as close, essential

---

[87] Subtle is the Lord, by Abraham Pais, Oxford, 1982, p141

[88] These repetitions are deliberate, driving some points home with a sledgehammer; for the new ideas about time are so strange that the supporting facts are repeated in the hope they'd make a difference, for otherwise people just laugh at any new ideas about time.

[89] The mathematical Theory of Relativity, Cambridge, 1930, Ch.1.

What is time?

and inseparable from life as water. But how does it (or did it) come about if not from the Heavens? The consensus in science is that it just is. Maybe, but even so how do we define it?

Our own Lord Bertrand Russell was the world's greatest philosopher living--- Logician, Mathematicians of genius, writer of genius, and as clever as, or more so than, Aristotle. For Aristotle had demerits, Russell had none, just pure brilliance; and as for science, he founded the philosophy of science with the book "Our Knowledge of The External World". He it was who defined post relativity (or secular) time thus: "It seems that the all-embracing time is a construction, like the all-embracing space. Physics itself has become conscious of this fact through the discussions connected with relativity."[90] Yet it failed to give us a clear philosophic idea of what time is and how it began as a secular entity. Of course people have been using time for thousands of years; it's only recently (since Einstein) that they're asked to define it differently than the religious or traditional view of it, and they don't like it; so they simply ignored what Russell said and continued to regard time either as just is, or divine, with some writers making fortunes upon fortunes for regarding it as linked to the Minkowski space to make it possible to travel by curved space---hence the popularity of time travel or the notion that time and space constitute one entity and the mystery

---

[90] Mysticism & Logic, George Allen & Unwin, 1917, Ch viii (x).

of time deepened. Otherwise Time Vs Space as a serious topic for publication would not be entertained in a science magazine.

WHAT A MOMENT MEANS

A moment is any sort of contact with nature, meaning all acts of perceiving (visual, tactile, etc.), so long as it is determinate and has to be repeated to continue---no matter how long it lasts. And that is the interesting point, because it can be divided and still be a moment or part of a moment.

At the time of Russell and Einstein Professor A.N Whitehead too was alive, and he was some brain. He defined time originating on this or any planet as post relativity time, implying that divine time could not exist. In The Principle of Relativity Professor Whitehead wrote: "...a moment of time is to be identified with an instantaneous spread of the apparent world"--- in other words, a moment of perception, vision, existence or 'Being', and went on, "...A time-system is a sequence of non-interacting moments [however that moment is defined]".[91] He was not a very lucid writer, but I suppose this is what he meant: every unit of time is a moment in life. Of course some of these moments are very long, like the year, or day, but they are moments in the sense that they are determinate units or

---

[91] The Principle of Relativity, Cambridge, 1922.

What is time?

periods that have got to be repeated to continue. This is an attempt to find a definition for time that suited its new status as a secular entity, created or constructed by ourselves----the year for instance, pared down to the seconds or even the atomic units of time which have always to be based on the second to make sense. Otherwise there is only one day and only one year; so any time system based on them is bound to be discrete.

This is where the problems began because discrete time cannot run all through the cosmos. Discrete time cannot bend, move or march because it does not flow through the universe; discrete time will not make time travel possible since it cannot flow forwards or backwards;[92] discrete time does not create the story of history, which is rather seen as the march of events not of time for only events can move forward, so that we carry the past with us always to the present and the future. Thus Einstein was right: past, present and future are mere linguistic illusions: you live with your past and will carry the same to your future---examples are everywhere, your bank balance for instance! You do not have to visit the past to access your

---

[92] Discrete time is spent and gone as an individual unit of time unconnected to any other---the years, for instance. Each year is determinate, unconnected to the next or the last. This is easily proved: one second before a New Year is still the old year; one second after is the New Year. All units of time obey the same rule because they are fractions of the year and can't spare or delay any units without eating into another's periods. This is a technical point in logic and mathematics but I hope it is clear enough for the reader to comprehend the argument.

bank balance; and the balance tomorrow will inevitably be what you have today----the bank manager will not pile pounds into your account for no good business reasons.

I honestly cannot imagine how anybody of whatever status, intellectually, academically, politically or religiously, could contradict the reasons for secular time sketched above to claim (by revelation or whatever) that time is other than the post relativity concept of it. Yet what keep appearing in books and magazines are still concepts of time in the old format: that it started from Time Zero, it just is; it increases, runs faster, slower and so on---yet discrete time cannot do any of these things; and since our time is based on the year every year, it is none other than discrete. I am often confused when scientists refer to something called "The Dawn of Time" unbroken to this day. And they are fond of demonstrating this by arithmetic as they count the years. Shouldn't it be "The Dawn of Existence"? According to all the authorities cited about time or post relativity time in this book, there appears to be no "unbroken streams" of time running all through the cosmos from the Dawn of Time or Time Zero---isn't that what relativity time means and isn't the contrary idea exactly what Professor Eddington called "meaningless noises"? Otherwise there would have to be different streams for the various units of time as we couldn't have one stream of time running as seconds, minutes, hours, days, months and years---quite impossible to program into a clock. Even the traditional definition of time acknowledges this. It speaks of "The passage of existence". In a multiple stream time it would be

What is time?

"The passages of existences"---plainly an illogical notion, as it would be a world of intermingling streams of periodicities all passing away at the same time with different velocities or momentum.

I must stress again that it's not right to say time does not exist, but nobody can define it logically other than as 'a construction' out of the features of the earth and other astronomical bodies. Sir Arthur Eddington was a very clever mathematician and scientist of genius, the founder of Astrophysics, and he says any such ideas about time after Einstein are fatuous---"meaningless noises", as he put it. That should keep the 'Doubting Thomases' quiet, at least for now. To help them along, I give below a brief sketch of the new theory---already I have shown that it solves the passage of time in a flash.

Now, currently professor Palle Yourgrau of the United States is making a name for himself as the champion of Time Travel, because he has written a new book called A World Without Time (Penguin, 2007), in which he claims that Time Travel is 'a scientific possibility', (which is evidence that the old theory of universal time running all through the cosmos is still prevalent in a great deal of the academic world). Time as a moment of existence (no matter how it is perceived) whose succession creates the illusion of continuous time can be incorporated into science and is backed by experimental results, Bertrand Russell, Professor Whitehead, Einstein, Professor Eddington and Gottfried Leibniz. However, time "as just is" (not given any definition), and which began at a date chosen by the writer (like Archbishop Ussher) is the bedrock of mysticism and has no place in science---yet scientists seem unaware of this. For fifty years my manuscripts are never even read before rejection. The reason, I fear, is that everybody wants to believe that time travel and life after death may be feasible with the mystical theory of time we have at present, so that life will go round and round the cosmos through deaths and rebirths forever. I am afraid we are doomed, not safe even in the hands of our own scientists, and Pythagoras is responsible. Professor Yourgrau's book, quoted above, which argues that time travel is 'a scientific possibility' has sold millions! I do not agree that life is brutish and sad---it is man who makes it so through greed, ignorance, pathological cruelty, religion, unreason, wickedness, fear of death, plus the seven sins. Considering the causes of all these evils, Sigmund Freud seems to me to have been the most rational thinker about social issues in all history. For when we are friendly, loving and kind, full of gaiety and generous to one another as at weddings, life is so sweet that we all want to live forever, but due to psychological defects in all of us, we cannot always be like that---for that reason, Freud is supreme.

Nevertheless, crucially, Professor Yourgrau has provided evidence that Einstein did not, in fact, even try to understand the Minkowski theory of four-dimensional continuum, or, in plain language, the theory that space and time constitute one entity: that Minkowski has linked (or equated) space to time.[93] He wrote, and I quote: "Every boy

What is time?

in the streets of Gottingen understands more about four-dimensional geometry than Einstein. Yet, in spite of that, Einstein did the work and not the mathematicians."[94] He himself quoted this gem from David Hilbert, therefore we can be certain it is true. Yet we know that Einstein used the Minkowski formula in his general relativity. The presumption then must be that he did so just to placate his mathematical critics who were calling for him to be hanged by the nearest lamp post.

Technically, it is quite impossible to equate space to time by means of mathematics unless one relies on 'i'; but then any such theory will fail to carry conviction because time is not imaginary and 'i' can only be used to represent imaginary quantities.

The creation of space-time, being the merging of space and time (still as separate entities), was achieved in special relativity, as Russell has observed, by the use of the 3+1 formula, but mathematicians were spitting blood because time had been made secular at the same time. They said it means man (as emotional, biased, partial and fraudulent as he can be!) creates his own time and then adds it to phenomena and call it objective reality. That's not an

---

[93] One has to admit that if this is true then the world has changed out of recognition, and time travel would be possible. The irony is that it is not correct yet the world has changed through Einstein's own theories of time, namely that time does not run through all nature and the same everywhere, and that our own time was constructed by man.

[94] Palle Yourgrau, A World Without Time, Penguin, 2007, p6.

acceptable concept to represent true reality, they claimed. Yet 4-D geometry does not and cannot exist, so the theory of curved space-time is flawed, and time travel via Minkowski is not feasible. The history of the Minkowski efforts is interesting even before we come to the story that Einstein did not understand it, for saying that in reference to any subject in physics or even science generally, means the great man thought whatever it is was nonsense. But Professor Eddington called 4-D Geometry arbitrary and fictitious (though useful for the study of phenomena in his Mathematical Theory of Relativity). Russell said it was compounded for the convenience of mathematicians (The Analysis of Matter). And one reference work (at least) called the theory artificial (The Routledge Concise Encyclopedia of Philosophy). It is true, of course, that Mathematicians adore the Minkowski theory because it enables them to dispense with the 3+1 formula, which they regard as less than objective for science.

In my judgment, I suspect Einstein did not even bother to understand 4-D geometry. I can imagine what was going through his mind. The mathematicians had to be placated and come to support his theories so that he could get on with his work. For that purpose he praised Minkowski and pretended to adopt his formula. The whole world was misled into thinking that, because Einstein praised Minkowski, the later had a hand in the theory of relativity. The truth is that he tried to contribute to it but failed because he had to base his theory on imaginary time coordinates. So, in fact, the Minkowski formula was

What is time?

irrelevant---there is no way it could have vitiated the new theory of gravity, and I would bet my last penny that Einstein knew that. The definition of time was a different issue. He said (and I quote from Abraham Pais's Subtle is The Lord...): "All that was needed was the insight that an auxiliary quantity introduced by H.A Lorentz, and denoted by him as local time can be defined as 'time', pure and simply". The many times this is quoted reflects the stubbornness of mathematicians!

Finally, in this section, as already mentioned, in the absence of a universal time, Russell demanded to know what is measured by the clock. In fact, there is nothing. So what is time? Nobody knows what time is, but by using the Lorentz concept that time can begin from anywhere, with Einstein's support that there are as many times as there are inertial frames, we can logically infer that time is constructed by man and that it is purely psychological and cannot exist outside the human mind. Mathematics is no help here. Generally mathematics is indispensable in physical matters not in matters of the mind.

Yet, in the end, time is basically counting cycles; therefore it's mostly psychological. Ten orbits of the sun are ten years. That's true, and in the absence of a universal time nobody can define time besides the mental notation of the orbits as time (the years of course are our years, the measure of our own ages); but crucially somebody must be there to set the points for the yearly cycle or there will be no years and seconds derived as fractions of the year; so it is also true that we create the years or time, as Russell has pointed out. The theory of the new concept of time is that any cyclical or regular motions divided by points will provide periodic intervals or time units for cultural use as the logic of time in the universe; that is how we get time to put in the clock. We can simplify this in mathematics thus: RM.P =TS + E (or RM.P+TSE), meaning any "Regular" "Motions" "Divided" with "Points" provide "Time Sequences" as the logic of time in the universe. Beyond that there is nothing for the clock to measure as time units. The 'E' represents "Existence": Regular Motions Divided with Points provide Time Sequences for defining Existence---defining, justifying, legalizing etc. The logical reason is that (culturally) everything in existence has to have its "when" (or time) of existence in the universe. For while there is no universal time, there is nevertheless a universal law for time and existence as the definition of life: every 'Being' has to exist or be covered by time so that it can be defined as existing at so-and-so a time or it never existed. Every existence can only be quoted in time, otherwise how can it be cited? This being so, the above equation becomes the equation by which all life can be defined or justified either in law, philosophy or science. That's how important time is, and we can say that it's deduced from Einstein's ideas.

It must, however, be noted that time is backed by, or based on, duration, the real unknown mystery of time, which enables us to tell that one hour is longer than one minute. The logic (or the law of time in the universe) arises from the fact that duration has got to be divided to obtain the culturally indispensable units of time, since all time is

What is time?

known and used only in units. The word time means nothing in culture without quantification---meaning 'the quantity of time' involved. Let's suppose that you are asked to sit down in one place for one hour as punishment; other persons are to do so for five minutes each. As you sit there people come and go rapidly every five minutes. You're ignorant of time units and their various durations, so you begin to wonder what is the difference between five minutes and one hour? In other words, how do we differentiate the lengths of duration between the different time units? In all the universe there must be a law or method for doing that, and, 'speculatively', I believe it is this: we count cycles based on the points used to divide the earth's orbit of the sun, and that is tantamount to counting shorter cycles with points out of the total duration or length of the orbit of the sun. We have no other methods for acquiring units of time; and this will be true of any determinate cycle used for reckoning time in any part of the universe. For the discrete nature of time is caused by the determinate cycle upon which it is based---year after year after year. There is only one year; we repeat it to get all the years we speak of. And since the year will end and restart we have to divide it into an exact number of units to coincide with the end of one year; this can only result in a time system that is essentially discrete with all the momentous implications of discrete time---something originating from and ending with man, which shows the nullity of all the prognostications of all the religions for a start! Time is not an easy matter; it controls everything. The irony is that it is 'constructed' by man, as Russell noted about a hundred years ago---we actually do create our time, though from factors naturally existing in the universe.

Duration (that is, during the period a thing is there or is encountered), of course, is natural and existing all over the universe and (under QED) it must be 'caused'; but to exist (or live in an inertial frame and have culture) you have got to find a mechanism for quantifying duration into time sequences, dividing it into manageable units. So duration caused by many factors (inertia, motion, atomic and nuclear processes, ebb and flow, force, obstruction etc.) must be existing all over the universe, but it takes the human mind to invent time sequences out of it. Luckily for scholars, for once it is plainly not a chicken-and-egg question since the factors, conditions and parameters for 'constructing' time sequences are all present throughout the cosmos. But since the construction requires the intellectual use of points, we had to come down from the trees to learn to do so. The year, for instance, is determined from one point to another: from midnight 31st December to the next midnight 31st December and then start another year on and on forever. Even one second before midnight is this year, one second after and it's the New Year---thus it must be recognized that the second is part of the yearly cycle. Intelligence or sentience is required in time's construction, plus a theory of numbers, the ability to count and arithmetic. Time does not exist outside the human mind, which, of course, means another line of inquiry must now begin!

What is time?

Thus time, as important as it is, is an artificial contraption based on existence for the justification and regulation of that existence and all activities. Let us say we could use the hand round and round without muscular strain. Ten cycles means it is time to go to school. At school, a hundred cycles means it is time to go and play, another hundred cycles and it is time to go back home---something like this suggestion (as a mechanism in a clock) can be seen as a device for reckoning time to regulate activities. Existence or Being seems inseparable from time in that everything we do can be ascribed to the expenditure of so much time; but that is not all that mysterious, because time is created by the critical and indispensable planetary conditions that sustain life, so that life cannot move without them, and as a result cannot move without the time they create for us. Like language time is secondary to life, something that helps us to know how well and safely to live on the planet. Take the daylight for example. All we need is something like the sundial, where the positions of the shadows of astronomical bodies (or whatever) enable people to know how to go about their various activities. This should not be interpreted as the motions of time; that it moves from morning, to mid-day and to the evening and nightfall---not at all. Only the shadows do move, but it has traditionally been assumed that the motions are those of time itself. That is wrong. Time does not move but repeats its units to pass by, as I keep repeating.

Rather time (or the time) consists of the units or portions of the earth's orbit of the sun we have predetermined in mathematical units and apply to events. Thus all time has astronomical roots, and the motion or advancement of time is achieved through the replication of its units: one hour means the second has been repeated 3,600 times, and so forth, not that the second has physically moved through space to become one hour. Time moves through replication not physically. Only the units multiply; the result is that man is confronted by two quandaries: (a) it means we can only know how time passes by and not what it is; and (b) people who cannot comprehend this regard it as one of the riddles of time that confirms their religious beliefs about the 'mystery', and this belief is so deep that nothing can dislodge it, which is sad. Because motion, events and everything else do not constitute time; rather we apply units of time to them; these units of time (or space) are obtained from a breakdown of the earth's journey round the sun, and mechanized as units of duration---seconds, minutes, hours and so forth, so that 31,536,000 seconds equal to exactly one orbit of the sun as 'a year'---every second is equal to a certain amount of space. The result is that duration is oppressive, because the earth's orbit is strictly fixed, repetitive and unstoppable. Otherwise there would be no duration and no time. That is why the universe is not regulated by time but only chance. Time is not the activity but the sense of duration (derived from the breakdown of the year) we apply to events, even including passive activity like sleep. However, tradition, religion and the sheer force of habit preclude understanding of this new theory of time that arose from the Lorentz discovery of local time or t^1,

What is time?

which Einstein cleverly interpreted as 'time, pure and simple' through his unique brainwave. I have been saying this for more than fifty years, but nobody will listen to me! Instead ordinary motion is regarded as time---yet activity is not time but what we use to determine its duration which comes from the sun. Another aspect of time is physical changes 'over time'. Changes that occur 'as time goes by' are not in any way related to time but occur through objects' own chemistry or other conditions like force and accidents. Such incidents happened even when we did not know of time; since then we have learnt to apply time to regulate accessible events as part of rational thought and science.

Actually the sundials give us the wrong impression. They show time as moving physically across the sky; in fact units of time are applied to successive motions of bodies. The seconds do not fly through space; rather we apply them to whatever we encounter whether in motion or not as stressed above. But the application of time takes the form of repetitions of its units: ten minutes means the second has been repeated 600 times. We may not notice it but in logic that is what it is---that is the metaphysical nature of time. A period of a thousand years starts and ends with a single second. Units of space have been converted to units of duration basically in seconds as our SI of time. The many categories of time (days, weeks, months and so forth) may be culturally useful but of no account in metaphysics.[95] Once the second is defined all units of time are covered since they are multiples of the second. Understanding the logical status of the second makes the essence and passage of time easy to comprehend. At least some of the legends, bogeys, myths and baseless mysteries of time due to religious beliefs can now be explained logically clearly for those who are capable of rational thought---indeed time can be demystified completely, though we know that not everybody will even agree to look at the evidence! And this brings me to a brief discussion of what role philosophy can play in all that.

I concluding this section of the book, I feel it is necessary to expand the notes about what happens on the planet being irrelevant in the cosmos at large. It seems obvious because the universe is rather so hideously vast and complex, yet it is still necessary to explain the point in some detail. In the future I expect the study of time to become scientific, why not? Economics and agriculture have qualified, why not time?[96] It has known attributes, it is

---

[95] For example, the passage of the days is not the passage of time because what happens on planets (billions of individual perspectives) cannot be relevant in the reckoning of what happens in the cosmos of so many billions of galaxies, and among whom even our mighty sun is like a firefly. The problem with time is that, while we look at it in human terms, its application is in metaphysical terms: it is what we need to live and therefore it is metaphysical; on the other hand we have no way of defining it except through human eyes, yet man does not count in the universe. Thus writers face great difficulties about time: what sounds logical is inhuman; and what is promoted as 'human' about time is inevitably religious nonsense.

What is time?

a vital subject or entity, we now know how it began or 'begins'---through the intellectual use of points; and therefore what it is made of, namely units of space, not that it just is; it has a quantity for study; we know it will end with the demise of the planet. We know how it is used or how to use it, and we know how it passes by---i.e. by means of replication and not physically through the air. For thousands of years man has been searching the Heavens for grand theories to solve the problem of the passage of time, while staring daily at the seconds repeating themselves to cause this same passage of time---ten minutes means the second has been repeated 600 times. There is no other logical way to account for the passage of time, I am afraid. There is only one second repeated to infinity, and it accounts for all time---the non-interacting moments of Professor Whitehead start (or started) with just one moment and is still the only time there is. Above all, the myths of divinity, fixed nature and total, absolute cover for the whole universe are gone. It's even foolhardy

---

[96] Once time is given the full scientific treatment researcher might like to look into what it is. At present they say it just is there; on the contrary I argue that it is obtained through moving from point to point and that without points instants could not exist. But to have a meaning to life here on earth, instants are called 'seconds' or that the basic instant is the second. The seconds are linked to the orbit of the sun so that a certain number would equal to a completed orbit. However the orbit is caused by gravity without which it could not occur. So does it mean gravity plays a role in the having (creation) of time? If so then it means it is involved in our existence too. I personally think gravity keeps our planet at a suitable orbit of the sun for water, vegetation and life to develop.

to believe that one time system could cover such a complex universe!

However, at present time is like the affairs of the human body as against events in interstellar space or the cosmos at large. Man is known to the earth and we can affect many of the natural phenomena on earth. The earth too is known to the cosmos and can affect many events in the cosmos through gravity; so the earth is relevant to the cosmos but not the little creatures crawling on its surface who are infinitely variable and in a continuous state of births and deaths, with infinite variations in uncountable body functions and numerous good and evil ideas floating constantly through their tiny skulls, and so forth---these are unknown to the cosmos and it certainly has no capacity to care about them.[97] Therefore the minutiae of our existence (including the time we have invented to regulate our lives) cannot be relevant to the cosmos or interstellar space in anyway. This is my definition of the metaphysical status of man and his time: Day&Night, months, weeks and so forth are all irrelevant. Only the yearly cycle counts because it is caused by gravity and therefore has effects on other bodies. But ordinary, humankind incidents on plants do not

---

[97] This is the perennial problem for mankind: we are completely alone in a world so impersonal and senseless that it has no idea we are even here, while the parent universe is so vast it can scarcely recognize the earth with a magnifying glass, let alone the tiny creatures crawling on its surface---religion too has failed us through the weaknesses of human nature. Thus man's metaphysical loneliness is complete. No wonder some people demand the right to die for one reason or another.

What is time?

count in the cosmos which is so vast that it can only deal with (feel the effects of) massive events. Nothing personal but an incident has to be massive to count or even cause a ripple due to the size of the universe. We are, of course, talking about hundreds of billions of stars, some of them so huge that millions not of people, not even of our earth, but of our sun would find room in them. Can the effects of our mental concepts of time reach them? If not, then how can we accurately estimate their age with our time? Who do we think we are just because we have time which the religions claim to be divine?[98] Take, for example, that a person decides to turn round and round perpetually (like the rotations of the earth), or like a fly flipping its tiny wings continuously, why should that mater to the billions of people on earth? Similarly, how could the earth's rotations matter in the cosmos of billions of stars---or how at all could they be noticed and by who or which of the billions of stars in the cosmos? Our time, being conceptual, is applicable to the earth only, exactly as Russell put it: "There is no longer a universal time...only the various proper times of the various bodies." Our existence on the surface of the earth is not felt even by the planet, and the planet itself, as

---

[98] I have always believed that because of time religious people think there is something important in religion; in fact there is not. Those with the contrary view should tell us where it came from. I agree that those intellectually unfortunate people who cannot live normal life without worship should be allowed to do what will help them, but only as their own personal thing. But the problem is that they don't want to do that; they rather want to slaughter unbelievers or convert them.

huge as it seems to us, is not felt in the universe except as a minute gravitational tag on some adjacent planets, comets, asteroids and the moon---not by the billions of mighty stars. So what we do and least of all the airy thoughts in our minds, like our concepts of time, do not, and cannot appear anywhere outside our bodies, let alone noticed by anybody outside our heads to have any effects on astronomical bodies. After all, concepts and thoughts are molecular and invisible to the human eye---a grain of sand is billions of times larger. The real mystery is why these molecular actions in our heads are able to give us ideas about the whole universe because of how the brain is organized, and further how following such ideas we can act rationally in our daily lives. However we are informed by scientists that due to the actions of electrons and photons, electronically atoms are capable of doing that from the brain (that they are doing it all the time in computers used to operate complex machinery), and that even the brain is not unique as these things can happen from anywhere if the atoms are properly organized---that trees or rocks can have minds, and it may be happening somewhere in the cosmos even now.

My definition of time as something we invent or construct to measure duration—the real mystery of time--can be inferred from the following scenario: let's say you are blind and you touch something (that is Professor Whitehead's 'moment' of perception or contact with the world). When asked 'for how long?' and simply because you are blind and cannot read the clock, you 'construct'

What is time?

your own time (Russell), and say, "I tapped my figure ten consecutive times". That's time without mythologies, arrows or divinity. Another mundane illustration is a search light, a powerful one shown on a particular spot, say, from a-search-and-rescue aircraft as happens all too frequently. If one wants to know 'for how long', tapping the finger will do---that also gives the length of time, without knowing what time is. That is the important point. We can only use some kind of repetitive motions to indicate the passage of time without knowing what the nature of time really is; we count the motions and say, wrongly, that that is time going or that it is the motion of time; in fact it is the cycles going, the motions of the cycles. If we want to provide an objective method to be easily copied and applied by all human beings, we use the obvious one, namely orbit of the sun that affects us all equally and therefore unlikely to be missed.

Chapter Two: The Nature and Origins of a Time System

We live and can only live according to our time system. This makes time second in importance only to life, no wonder the religions insist on their own interpretations. All through human history, and beginning with the Day & Night system (sleeping at night and working during the day), we have lived according to our notions of time. It's so oppressive that in many cases we simply have no choice. Since all calculations of time are based on the motions of the earth, and as we can only live by time, it means we live and can only live by the conditions of the planet, as a result we are compelled to live according to the planetary conditions; it is virtually unavoidable; and also miraculously or providentially it ties in (scientifically) with the idea of gravity citing our planet in a habitable zone in the universe beneficial to human existence. It is a religious as well as a scientific notion, but I am obliged, as an honest writer, to point it out whatever may be my personal beliefs. The time units are mathematically based on the orbits of the sun by the earth as fractions thereof (being products of points), and since the orbits are repetitive and unstoppable, the units of time pass by oppressively, giving man no choice but to abide by them in other to live safely on the planet. This situation is, as stated above, the origin of the sense of duration, the oppression of time and the reason time waits for nobody--- frankly it can't! For nobody can stop the earth. But it makes time look like a divinely imposed force and some unscrupulous people have claimed it is. However, in logic, the explanation given above is the cause, not God.

Again, everything is related to, or based on, our time system. For example, the whole of the quantum theory is based on time, but on whose time, since there is no longer a universal or cosmic time? This is an ancient problem in philosophy, and we're not even ready to face it yet. We think the quantum is the 'smallest bit of matter that can exist'. Yet it is time-dependent: so much energy emanates only after the lapse of so much time---I am trying to avoid the technicalities that put readers off.

The whole of our life is dominated by quantum mechanics as we learn from QED. But in the absence of a universal or cosmic time, whose or what time underlies all this? How does it get mechanized in the clock? The answer is rational time, the logic of time, or logic and time, otherwise known as 'the quantity of time' or quantified time units, like the year pared down to the seconds. It's the only way to acquire logically structured time without which there can be no cultural improvements. Before that insight we simply did not have a clue what time was. However the repetitive cycles we call time (say, the years) are merely physical events. They can only show how many cycles (years) have passed. If we translate that to 'how much time?' it means we can only know how time is passing and not what it really is.

Now, if we ignore Archbishop Usher's infantile, religious time of Creation as we must, then there have been four intellectually respectable time systems in the world overall. The first was just a process of people marking signs on the wall to show days and nights as they passed by or succeeded one another. It included the Sundials and many other primitive practices. (The concept of time 'passing by' is basic to all systems of time ever invented.) It ended with the Newtonian Absolute time, as our second system of time more intellectually respectable than any system before it. Then Lorentz discovered time dilation through physical experiments implying that absolute time did not exist; he called it 'Local Time', or t_1, and refused to believe that it was normal time. Later he confessed that he might have been able to discover special relativity if he had paid due attention to his discovery---which Einstein was clever enough to pounce on to work out his theory of frames and special relativity, et al.

What is time?

The fourth system of time was the Minkowski four-dimensional time, or 4-D geometry, which was purely mathematical. Everybody knows how cheekily he introduced his contribution in his Raum und Zeit (Space and Time) lecture in Cologne, 21 Sept. 1908: "The views of space and time which I wish to lay before you have sprung from the soil of experimental physics, and therein lies their strength. They are radical. Henceforth space by itself, and time by itself, are doomed to fade away into mere shadows, and only a kind of union of the two will preserve an independent reality." We're all taken in by Minkowski's confident proposal, including Einstein, or so it seemed. Foolishly, I once wrote a small book entitled The Mathematical Theory of Time based on the same Minkowski theory and then withdrew it immediately before philosophers of science had time to send the men in white coats after me. Not a single copy was sold because I realized that it's rubbish or based on an illogical proposition. But at the time everybody thought Minkowski was a breath of fresh air.

According to the brilliant British science writer, Dr John

"Minkowski's geometrical description undoubtedly improved the clarity of the special theory and is still regarded as the best way to understand it." Well, if so then I am not surprised that some writers are saying even now that relativity is not properly understood. Set that against what Professor Bernstein wrote: "In the absence of gravity space and time are distinct entities. In the metric of special relativity they play distinctive roles..." Time can be merged with space but by the 3+ 1 formula not by mathematics alone---for it all depends on how time is defined. Minkowski did not define it but obviously his theory implied that it is something in general existence. On the other hand, my suggestion is that, as Russell was clever enough to note about a hundred years ago, under relativity time is 'constructed' by us. The elements for doing this construction may be everywhere in the universe; but it takes human intelligence to do the construction. So my inference is that the universe is not regulated (or governed) by time; it requires logical sequences, but I doubt that outside the human mind logic does exist. It seems to me that every event in the cosmos occurs through chance, accidents or chemistry. Logically structured existence requires a mind and the kind of stupendous mind to control the cosmos is scarcely imaginable---to manage a hundred billion stars in one galaxy?

Since there is no longer a universal time, there is bound to be a method for acquiring time, or what we use to guide our actions and which we know as time sequences. This method is what I am in the habit of referring to as 'the logic of time in the universe', in the sense that any 'Beings' in any part of the universe will have to invent a time system by the same method; and I imagine that the method or logic will include the following line of thought. Nobody can ever know the true nature of time. However, we do have intervals between events and therefore also between points; they amount to 'periods of waiting' in the mind. We then employ repetitive cycles or motions to give us culturally useful units of these periods and apply them to events as units of time, and as the time units are conducive to the earth's condition, we are then able to live safely on the planet. In logic moving from point to point gives us the passage of time---which is all we can ever know of time and which we can even mechanize in a clock due to the built-in logical sequences. It means using any cyclical motions and counting them as the units of time to guide any action that is not continuous like motion on a conveyor belt. This is the logic of time in the universe---tapping the finger amounts to the same thing in crude terms, but it shows that time is not that mysterious. Yes it is complex, but so is language.

Chapter Three: The Minkowski Equation of Space to Time

In a short digression to consider what Dr Gribbin has stated above, I would like to comment on relativity and the mathematicians' interpretation, regardless of what Einstein has said about the matter. I have received several dissertations about Minkowski and his equation of space to time. In one of these, the writer wrote the following in his attempt to explain the gist of the Minkowski formula: "Minkowski's space-time is his way of interpreting Einstein's Esynched system...Esynched clocks obey Lorentz's

What is time?

local time equation...in which t is the general time... and because the 'time' per consecutive 3d point is different than that of its fellows in the direction of motion, the 'time' in that system has a continuous range from infinite past to infinite future. That is the cause and meaning of Minkowski's 'four dimensional space-time continuum". You'd need ten PhDs to understand this, and if you do, then you'd have to inquire from the writer what is measured by the clock as time for general use? For example, the writer mentioned 'Space-Time' yet he couldn't define it clearly. Here is how the great mathematician and logician, Bertrand Russell, defined or explained space-time for everybody to understand what it means: "SPACE TIME, as it appears in mathematical physics, is obviously an artifact, i.e. a structure in which materials found in the world are compounded in such a manner as to be convenient for the mathematician." You won't need a single 'O-Level' to understand this, meaning that it is not true of the external world and so those mathematicians regarding it as such are willfully distorting the theory of relativity and mathematical physics as a whole, 'willfully' because they know it's logically untenable. A word of advice to inexperienced mathematicians: philosophers take all mathematics into consideration before deciding on the nature of the external world. So the bizarre notions dreamt up in mathematics are most unlikely to render existing theories valueless, but they can be consistent with them---improvements are welcome, but revolutionary ideas from mathematics are not common. That, precisely, is the problem between Einstein and those mathematicians yearning to prove him wrong or grab a piece of his fame. Einstein's ideas constitute physical intuitions or insights about the whole universe. You can apply mathematics to them; but you can never undermine them with mathematics alone---the physical constituents would have to change for that to become possible.

All those (philosophers, logicians and physicists) who think deeply about relativity will get a sharp pain in the mind that (defined like this, as our anonymous writer has done above), relativity will never be understood by those with professional obligation to write for their fellow mathematicians' understanding of the theory, which is completely wrong and contrary to the philosophical interpretation of Einstein's ideas. Russell said space-time is artificial and compounded for the convenience of mathematicians. That's not an honest search for the truth, especially when the concept involved leads to a view of the world that's the complete opposite of physical reality--- time is not the same thing as space, because the equation to make it so was not successful, as Russell has confirmed. Therefore space-time, used in the sense that space is the same thing as time and vice versa, is a distortion of relativity. Einstein did not just write on physics. He reconstructed the whole of physical reality. A hundred years have gone and we still can't understand his ideas properly because writers are picking what they can understand. Thinking about the whole of physical reality is beyond ordinary writers no matter their status in the universities---it requires a team of great philosophers to interpret him properly. And where are they? We have had

What is time?

but only one great philosopher, Bertrand Russell, but he's dead. It's most unlikely we'll get another one because the world of learning has changed: nobody is encouraging such thinkers; even publisher complain they have no expertise in these matters, and rather prefer to heap millions on convicted murderers to tell their stories for commercial gain. In the past it's different. We had publishers with genuine love for knowledge; today there is only genuine love for money, even promoting tarnished sources of money with criminals and murderers. People regard Einstein as just another scientist. That's a mistake. In all history he is the only man to discover how to think logically about the world. To equate energy to mass alone made him the greatest thinker of all time in my estimation; yet there was more, much more.

There are several specialists in science. But relativity is different because it is a logically deductive system of thought about the external world as sketched below. Of course the one theory will not be able to answer all of the countless queries about nature; no theory can do that; but for those professionally bound to work with it, I suppose they can only understand the theory's contribution by thinking about it as a logically deductive system, beginning with the fact that there is no longer a universal (or general) time due to the discovery of 'local time'; that is the genesis of the theory of frames and it is upon the theory of frames special relativity is built; and upon special relativity that general relativity is deduced. Einstein divided the universe into two: one is where life is feasible; the other is the metric of general relativity where life is not feasible, and he even went on to sketch the physics applicable to each half more or less perfectly. That is the caliber of the man whose ideas are handled so carelessly by mathematicians because they think they know best, if so why couldn't they discover relativity? Because of time, a great deal of humility is required in the study of relativity, or Einstein's ideas as a whole. Placed under philosophical scrutiny, the new concept of time, properly understood---without the Minkowski fiction---changes reality, human existence, philosophy and how we see ourselves completely, simply because time controls everything. For if time is not eternal or divine, then who are we since we depend entirely on time? Yet we have to accept that life and time will eventually end with the planet's demise

And there simply is no general time that can be manipulated with mathematics to "range from infinite past to infinite future". If mathematicians want to preach religion they should go to the churches. No wonder the Encyclopedia Britannica (Macro) states that almost all cosmologists agree that space-time is infinite in its timelike directions ---what will happen to this time when the earth ceases to exist? What will the word 'infinite' mean? Are we to suppose that it will fly out of the window to the universe singing the praises of Hermann Minkowski? Or perhaps time will continue in people's heads as they travelled round and round the universe looking for a new home as Pythagoras supposed? I accept that life is harsh, sad and brutish as well as being short, as the poets have often told

What is time?

us, and death is simply unspeakable, especially when one thinks of those left behind who're vulnerable. Unfortunately that's the curse of life, and yet we never ask to come to life; it's always someone's decision to bring us to life. The law recognizes this and makes our parents responsible for us, but nobody can go beyond that in law. Sad, but that's it. The only consolation is that death is not painful. We usually prevent human suffering, for that is what is painful in life---poverty, want, misery, disease, loneliness---worse of all mental illness. But in death we're not going to be cursed to look back to see what we're missing. Since the era of Pythagoras men have sought various means of coming back to life after death. It's no use; human life is not worth it. Life is painful, not death. I would advice everybody to try and live a good, trouble-free life, prepare for a decent burial and let his or her work in life, whatever it is, be good and serve as the lasting memory of him or her, so that those left behind could be proud of it, or probably benefit from it as well. Sadly here we are; people disparage philosophy and yet praise the distortions of relativity Minkowski propose, perhaps because only the philosophers could see that his theory was arbitrary and logically flawed.

Precisely what is wrong with the Minkowski equation of space to time is this: there is no time at all in nature, least of all general time. Yet we have something we use as time---how did that come about? This is the most important philosophical question since Plato, and Bertrand Russell put it in the clearest, lucid English words for everybody to understand, if he wants to understand it, instead of propounding mere mathematical formulas to please himself. Those who disparage philosophy should realize that even Einstein adored the title "Philosopher/Scientist, and those who do not know any philosophy or how philosophers think should stick to those aspects of relativity relevant to their fields and leave the wider and deeper interpretation of the theory to philosophers. Luckily, thanks to the Russell and Whitehead examples and teachings, many other philosophers are also accomplished mathematicians. These 'accomplished' theorists have also read the judgment of Professor Eddington regarding the Minkowski proposals---i.e. to the effect that they are arbitrary and fictitious simply because they are based on imaginary time coordinates for the creation of the mathematician's imaginary infinite time forwards and backwards.

There is no general time; and if there is no general time covering the whole universe then it means there can only be something called your own 'local time', this inference is based on the fact that we know, use and call something 'time', without which we could not exist at all because we'd not know the world properly---day and night system, the seasons, knowing when it is safe to go to the farm, when to sleep and when to wake up, go to school, work, the shops, and so forth. We have time. How did we get it in the absence of a universal time? Philosophical reasoning is the greatest achievement of the human mind, not mathematics. Mathematics is like logic, somebody has to weld their pieces into a philosophic whole.

Once local time was discovered proving that time is neither universal nor fixed for all eternity, ironically, the concept of local time ceased to be important---only the idea mattered. Man was freed from the constraints of absolute time, but then what is measured by the clock? It took a philosopher to ask that question, and it still hasn't been adequately resolved. My efforts (which are everywhere ignored because people still think there is general time) are part of the attempts to answer that crucial Russellian question. One answer is that time is variable and the variability is caused by means of unique parameters---that is to say, the elements (agents, factors, etc.) we use to create our local time vary from one place to another. Obviously the planetary orbits vary, so the years of the planets also vary---simple. But logically time remains 'intervals between points', since instants can only be produced by moving from point to point. Sentience is required.

In a fragmented universe, as the Einstein theory of frames makes clear, one system of time cannot be applied with equal validity to all fragments of the universe. But then what is measured by the clock? This Russellian question will never go away. Time Dilation did not dilate any time at all. The Lorentz local time idea was important because it inspired Einstein to discover his theory of frames. After that we've to discover how our own time is created---in the absence of a universal time, of course. This is what mathematicians are unable to do, but they will not accept the philosophers' solutions because they still believe that time is general, and use their esynched systems to create infinite time from the distant past to the infinite future, because mathematicians are telling us that time has been equated with space; for me this is precisely the reason why the Minkowski theory is a distortion of relativity

It is not science; it is day dreaming or religious nonsense. And to say that it is the only means to understand relativity amounts to admitting that Einstein's ideas are still not properly understood. My advice is to think of special relativity and the two postulates underlying it. Next, you should go on to consider the theory of frames upon which special relativity is based (with every frame having its own natural laws, hence the relevant postulates); then move on to the metric of general relativity that is another world or segment of the universe altogether; and finally ask yourself who on earth could have conceive that idea about the divisions in the universe showing where life is possible and where it is not? He even suggested some of the natural laws too—e.g. the bending of light, etc. All these even before we come to the quantum theory! At the very least mankind should show gratitude by understanding what he said and meant, not distort his ideas with meaningless mathematics. I agree that the cause of our having of time was not considered in relativity simply because time was regarded as just is, as I am struggling to point out. He should have shown how time would be constructed in any frame. But he did his best, which all anybody can do.

THE MERGING OF SPACE WITH TIME

This is where Minkowski is reputed to have shown his mathematical genius in the interpretation of relativity, making it accessible to scientists as a whole. But of course we know now that his formula was wrong because time is not mathematical. It is something else but can be rendered mathematical in presentation---and in presentation only. Several years later I came across the report that Einstein actually never understood the theory, although craftily he praised Minkowski to placate his mathematical critics at the time. According to David Hilbert: "Every boy in the streets of Gottingen understands more about four-dimensional geometry than Einstein. Yet, in spite of that, Einstein did the work and not the mathematicians."

I have gradually come to the conclusions that when something like this is said it means the great man thought the theory concerned was nonsense. Hence, logically what can be sustained is the concept of time proposed by Einstein (as mentioned above), in which time is seen as it is----year after year after year, or second, second, second and so forth. Bertrand Russell interpreted it as "relation between points", or that we construct time units as relation between points passing by---the years, for instance. And Professor A. N. Whitehead said it means a time system is "a sequence of non-interacting moments". Of course we have already noted that the Minkowski 4-D geometry or four-dimensional space in which space and time are fused into one entity was described by Professor Eddington as arbitrary and fictitious. Scientists continue to call it artificial and yet repeat the phrase 'space-time' to mean that space and time constitute one entity as Minkowski proposed.

My belief is that this is the main reason physics is wobbling. In my view, the phrase 'space-time' cannot and should not be used to mean that space and time constitute one entity; but it can rather be used to imply that time, as relation between points, can only be had by the application of points to space, for using points involves space---so you cannot have time without space, but the two are separate, precisely as Einstein made them in the special theory of relativity---see more of my arguments below. There are a lot of repetitions here, but I take the view that it is a small price to pay for clarity in a subject so mysterious that many writers are afraid to touch it! And this theory of time, as tentative and tremulously I put it forward, is nevertheless aimed at helping us to get an idea of how time is passing by; they call it the most intractable aspect of time, but to me it is all we can ever know of time. I don't think the passage of time is intractable; it's time itself that's intractable because it is impossible to know what it is at all; what we call time is only how it is passing by---year after year after year. Yet the year does not help us to know what the nature of time that is driving it really is. The year is only a physical journey round the sun. Thus I conclude that all we can ever know of time is how it is passing by and never what it is.

What is time?

Alternatively, we have to trace the metaphysical origins of time to know how it can pass by. We can't know that by assuming that it just is, for even so in what does it consist? It just is, as what? As time?---then we have to rely on logic to tell how it can pass by and that is by replication of its units, since time is known and used only in units. What ignorant people call 'silent time' is neither traceable nor usable. We can say that even when asleep time is passing---yes, because the earth never stops and therefore is continually churning out the seconds, meaning time by means of points and therefore accountable in logic. Time by means of points (the point-divisibility of space), linked to the orbit of the sun, seems to impose the sense of duration that gives meaning to units of time as pieces of space; but the mechanism is not yet properly understood---or I should frankly admit that the idea is mine but it has not yet been scrutinized or tested. Philosophy has changed since Bertrand Russell: we don't just say things any more, but only what sounds logically true; that is why he dismissed Wittgenstein as trying to put an end to physics; for these days no one can say much that makes sense unless he understood mathematics and physics. The world of sense is not an inference but a construction.

Perhaps it is necessary to lay out the logical objection to the Minkowski proposal. I will do my best. To be frank, it is not really important. Minkowski probably wanted a piece of the Einstein fame but he chose the wrong subject; yet, sadly, everywhere we hear of the term 'space-time' used in the sense that space and time are unified into one entity, and it's the theory of Minkowski they're referring to even though every Reference Work describes space-time as artificial, and rightly so.

The Minkowski theory is purely mathematical in presentation; but every mathematical theory has to have a logical foundation, and so it is the logic of the theory that is regarded as defective. We start with his replacement of time with an equation usually referred to as 'ict equation'. This is how Einstein himself stated it: "...we must [really must?] replace the usual time coordinate t by an imaginary magnitude $\sqrt{-1}.ct$ proportional to it..." That's enough. But even more damaging is what he wrote in another section of the book (RELATIVITY): "...the world of physical phenomena which was briefly called 'world' by Minkowski is naturally four-dimensional [note that so far he's given no logical reason why it should be so but soon he did so] in the space-time sense. For it is composed of individual events, each of which is described by four numbers, namely, three co-ordinates x,y,z and a time co-ordinate, the time-value t. The 'world' is in this sense also a continuum; for to every event there are as many 'neighboring' events (realized or at least thinkable) as we care to choose..." No wonder that David Hilbert, who must know, says Einstein did not understand or accept the Minkowski theory. He sure could not have overlooked the logical defect in the theory even if he was asleep, but at the time he needed the support of the mathematicians with 'superfluous learnedness'; so it is possible that he felt coerced, forced to used the Minkowski formula in the field equations of his general relativity,

What is time?

knowing that it could not in any way vitiate his proposals about the causes of gravity in the cosmos at large. I am really annoyed that physicists do not understand this; it's so simple. If I can do this without even primary school education, surely anybody can. The need is to think philosophically and they don't want to do so due to inherent professional bias against philosophy and philosophers.

The point is, the Minkowski theory is entirely based on Coordinate geometry. It had to be because it was supposed to make time geometrical; coordinate geometry has to have a premise, a logical foundation---it should be rooted in natural phenomena. The basic theoretical formula is often referred to as "The Transformation of Coordinates", or "The Lorentz Transformation". But at some stage the transformation become tenuous, even ghostly; yet one cannot logically discuss time in ghostly terms.

Time is peculiar and very strange; the most mysterious thing in the universe. It is so strange that several books, including the one mentioned above by Professor Yourgrau, called A World Without Time (note the title well), are saying it does not exist at all. Yet we have time: In sports, in work (you work according to time); in schools, in lectures, in travel, the trains and planes are all scheduled according to time. In everything we humans do time is the most important aspect. And if you are going to discuss time in the form of the transformation of coordinates you must realize that (in the nature of mathematics), at some stage the transformation will become ghostly. Even Einstein used the phrase "realized or at least thinkable". But it will not do. You can't use the term 'thinkable' in serious science or what he often referred to as 'logical thought', and he knew it.

I come now to the elaboration of my arguments against the theory promised above: a little theory here will help the reader to understand what I mean. Bertrand Russell noted that the merging of space and time (which Minkowski said he could make geometrical), is already evident in special relativity. It is so important that he called the Einstein theory of time perhaps his most revolutionary and far-reaching discovery, and I have rubbed the cautious term 'perhaps' to insist that it is in fact his greatest discovery far above his new theory of gravity. Funny enough it leads to the view that there is no such thing as a universal time, yet if that is so then the Minkowski theory of time which Einstein was forced to use for general relativity is completely untenable. For if universal time does not exist then space is not the same thing as time because space is universal no matter how it is defined.

Here is the reason why. According to Russell the merging of space and time is already implied in the special theory of relativity, and I agree with him. So why did Minkowski feel that he had to do it all over again? It's because he said he could render the whole exercise mathematical or geometrical---he claimed that time, as proposed by Einstein, is naturally geometrical so that we could dispense with the 3+1 formula. You cannot blame the mathematicians for falling in love with this proposal. It sounds fantastic and adorable---provided he could bring it off.

Well, was he successful? The answer must be no because everybody calls his space-time concept artificial, and although we continue to use the term 'space-time', Professor Eddington has warned everybody not to forget that it is arbitrary and fictitious, so let us look at the reasons for the Professor's advice.

Logically the merging of space and time means two things, or implies two conditions. The first is that you cannot have space without time, and the second is that you cannot have time without space.

First, the space: time is gone or going even as you traverse the most infinitesimal space; so every movement or activity must include time coordinates---a time coordinate is obligatory in everything we do, practically everything. Thus Russell pointed out in The Analysis of Matter that the tram never repeats a former journey; the reason is that, although running on the same lines, it is really in a different position because its time coordinates had changed. In physics it is even a different thing altogether.

The second point is that you cannot have time without space, because in the absence of a universal time you only get the time as relation between points—just like the way we get our basic unit of time which is the earth-year. We only can get time intervals as relation between points. The mention of points implies the inevitable use of space. After that space and time are separate entities, and that is how Einstein made them in the special theory of relativity, again as Professor Bernstein put it, "In the absence of gravity, space and time are distinct entities. In the metric of special relativity they play distinctive roles..." I am therefore not in the least surprised that David Hilbert has confirmed that Einstein didn't even try to understand the Minkowski theory of four-dimensional continuum. The extension of the time coordinates to make it logically valid would be infinite, in other words universal, yet he had already shown that universal time did not exist.

Perhaps Minkowski saw in the whole Einstein proposal the opportunity to make a name for himself as a great mathematician ---but obviously not a great philosopher of physical reality; the two subjects can never be treated adequately in separation. He proceeded along a mathematical route for making space and time one entity and said so publicly---hence the word 'space-time' and some writers have said Einstein was lucky to have Minkowski on his side. It is also true that he praised his great mathematical knowledge----but not his philosophy, I am afraid. However you look at it, the absence of a universal time is blocking the way to a logically valid proposal to make space and time into one entity, and a man like Einstein couldn't have missed it.

Because Minkowski based everything on the transformation of coordinates, his proposal is not logically valid since he had to rely on imaginary time or i in the discussion of time, or his basic equation usually called ict

What is time?

equation: $\sqrt{-1}.ct$... But, as I keep stressing, at some stage you cannot transform coordinates indefinitely, or to infinity because there is no longer a universal time---which is what the same Einstein had discovered. You can do so in mathematics---which is what he did--- but for an elusive thing like time (elusive, basic and essentially unavoidable) at some stage you are going to have to assume that time will always be there---but in the absence of a universal time, how can that be guaranteed in logic? How is the time going to be there, and there where? Or how? Time is always there; what we want to know is how it gets there, as Russell put it, "if cosmic time is abandoned?" Life came first before time; but once you're alive you exist in "a when?" situation. You cannot say somebody exist without knowing when? For nobody could be in existence without a time of his being. To live is to expend time.

Paradoxically part of the mystery of time is that this time has only gradually evolved as we learned to 'construct' it out of the parameters that were always there---we always aged, but have only recently realized the reason why? The same thing applies to time: arithmetic, a theory of numbers, sentience, and the ability to count had to be learnt before time could be constructed for general use in society. For centuries, man thought time came from somewhere, say, the cosmos; but thanks to Lorentz and Einstein, we now know that it is a concept (or a contraption) we construct from existing parameters a concept we form from percepts. But once that is done, existence has to be interpreted in terms of time---ironical maybe, but that's the truth. With this explanation of time, it ceases to be scary and all theories that makes it mysterious or universal (like something eternally existing everywhere and the same all over, including the concept of four-dimensional continuum from our mathematical friends), have to be set against the logic of time as we experience it. I can't put it stronger than that but the reader will know what I mean, instinctively.

Thus, so long as we all agree that there is no longer a universal time, the Minkowski equation of space to time (or the creation of four-dimensional space continuum) will remain invalid in logic---but not in mathematics, which brings in an interesting contrast between mathematicians and logicians or philosophers regarding the nature of physical reality. Like their patron, Pythagoras, Mathematicians believe that what is logically valid in mathematical deductions is or can be encountered in physical reality. That is called 'inference', but it's no longer applicable. Russell and Whitehead have shown that the world of sense is a construction not an inference.

Conversely, philosophers believe or think they know that what is proposed in mathematics must be discovered by philosophers and theoretical physics or is mere speculation, if you want to be friendly and charitable, or mere dreams from pure mathematicians as they are wont to do, if you want to call a spade a spade. Some very unkind things have been voiced about the habit of pure mathematicians, but then we should remember that even Newton said he's afraid of them---several of my own

What is time?

technical papers have been rejected by mathematicians because of my criticism of Minkowski, yet we are now coming to the conclusion that his equation of space to time is logically untenable. Who knows what progress we could have made if this was recognized earlier?

As stated above, it is possible that the Minkowski theory is causing part or even most of the problems facing physics at the moment; and we ought to realize that physics is now so complex and vast that no one scholar can stand up against it. With all mathematicians inevitably backing the physics establishment, it's impossible. Yet even Eddington called the Minkowski proposal arbitrary, so did Russell, and now we know that the great man himself, Albert incomparable Einstein, also did not accepted it---not even bothering to understand it. I accept this as true, since we know that he intensely disliked the mathematical interpretations of his ideas. Why mathematicians believe they hold the key to the truth in the interpretation of the world beats my understanding. Logic is the arbiter of the truth and is based on language not mathematics. There are several fields where mathematics do not apply; and even in physics concepts come first (as Einstein has demonstrated.) Concepts come to the brightest among us as 'Brain waves', or intuitions, before they can be interpreted in mathematics.

The Minkowski mathematical theory of time is a serious matter and I hope physicists will reconsider their slavish use of the term 'space-time' as meaning time and space constitute one entity. Space-time should mean the

3+1 formula only. Eddington called the Minkowski proposal arbitrary and fictitious but added, "...Such a mesh-system is of great utility and convenience in describing phenomena, and we shall continue to employ it but we must endeavor not to lose sight of its fictitious and arbitrary nature." I believe Eddington was partly responsible for the mistake over Minkowski, for this is a very learned ('British') way of committing a literary and intellectual crime of the century---urging scholars to continue to use a theory so roundly condemned by himself? And his offence is made worse by the cunning manner he pointedly refused to mention Minkowski by name. Nobody would mention this, but there were two theorists working on relativity---Einstein the originator and Minkowski the interpreter. This was known all over the scientific world. Professor A.N Whitehead pointed out in The Principle of Relativity that we owed relativity to both. So it didn't take me long to figure out that it wasn't Einstein's theory Eddington was describing but that of the other contributor to relativity!

Technically this is all because you can never extend the transformation of coordinates to infinity; so our very able physicists should understand this anomaly at the heart of physics. When this has something to do with time, you have got to define it first and nobody can ever define time. What we call the definition of time is only how it is passing by and never what it is. In the past this was not generally appreciated, as everybody continued to treat time as if it were a universal entity; but nobody can hide behind that lame excuse any more.

What is time?

Having dismissed the other three systems of time as logically flawed, even though people still believe in them, we are left with the true and pragmatic concept of time (the way we get the years), as one in which we 'construct' time, as Russell put it. That is how we can logically acquire the 'logic of time in the universe', showing how time (as non-interacting moments or time in units created as relation between points) is possible in any part of the universe, namely we simply do not know what time is, but can use repetitive cycles (like the year) to give us numerical quantities of how it is passing by, ten orbits of the sun means ten years have passed by and gone; and that, of course, can be mechanized into a clock as we are doing; it is the theoretical explanation of what we already have, know and use as time---no matter how we got it. Even religion played a part. In the words of Bertrand Russell (to quote him in full):"It seems that the one all-embracing time is a construction, like the one all-embracing space. Physics itself has become conscious of this fact through the discussions connected with relativity." The rest is a simple matter of straightforward deductions. However, it has to be acknowledged that human beings are extremely fragile both in mind and body (psychologically and physically.)

We should not really be here at all and we know that many factors have contributed to our survival, including religion. Some obvious myths may be mocked today, and certainly not wise to continue to observe them---human sacrifice, for instance. Yet in the dark old days they might have contributed to our survival. The same story underlies the having of time, since it is not cosmically imposed. That is how important the Einstein notion of time as a purely secular or psychological entity happens to be. Religion, astronomy and astrology, mechanics, mathematics, the ability to count, physical suppositions or concepts (as physics) and sheer human ingenuity or sentience, all contributed to human beings acquiring the concept of time; they are all part of the logic of time in any part of the universe. Sentience was particularly important because somebody had to be there to set the points for the yearly cycle. Any 'Beings' anywhere will face the same problem. This is what we can logically trace about time overall. And I have to emphasize that, despite the Minkowski contribution, which I call 'a distortion of relativity', Einstein in fact made space and time separate in the special theory of relativity.

There is something to be said for the notion that time does not exist. I have even written a small book about this problem, entitled Why Time is not a natural Phenomenon. But there is something we know and use as time. So time is real enough, but to be really philosophically or logically accurate, we should regard time as a device we have invented to guide our actions and that it does not exist naturally in nature outside the human mind. Thus time can be defined as: "existence" and does not move; what is moving up and down are the cycles we use to regulate our actions and which we call 'time'! This may be a paradox but no longer mysterious since Einstein's demolition of Absolute Time. So the phrase 'passage of existence' is

What is time?

almost right, except that time does not move, only the cycles we use to reckon how time is passing are moving, and the passage of existence is chemical not temporal.

Chapter Four: The Four Agents That Cause Time Intervals

I remind the reader that the concept of time upon which my analysis is based is that time does not run through the universe. Rather time is one moment and gone, as defined by Professor Whitehead. To have time to apply to continuing events we replicate the units of time as created with repetitive cycles like the year and similar cycles: thus we can have minutes, hours, days and years to apply to any events.

What is time?

Now, one universal constant of 'Being' is that everything in nature is caused, whether the causative agent and processes are ordinarily experienced by man or not. It is part of the duty of professional thinkers to investigate the causes of things and events as their full-time job. We are in the habit of calling thinkers philosophers; but there are other thinkers who are not philosophers---scientists, mathematicians, logicians, all of these work, in the main, on our problems, trying to solve them for us, and one thing they all agree on is that everything has to have a cause. What we call 'research' is the act of searching for them.

For ever trying to create problems for science, the religions go further to insist that man's life must also have been caused, and as we cause human births, somebody must have set the wheel in motion for giving birth to humankind or we could not have come to exist from thin air. From this they deduce that God must exist, and I would advise people to desist from trying to condemn or defeat them in arguments over this because they would rather die than accept any other (scientific, logical) explanation for the being of human life, since by this same doctrine death is only a renewal of life, and yet, despite their pretences, everybody is afraid of death; and of course nobody should try to deny people the right to believe what they will because some people couldn't live without their religious beliefs. Another thing we all agree on is that the causative agents may not be immediately apparent, their effects may delay through inertia or chemical processes, but causes are the agents (creators) of everything, gases, solids, atom, chemistry, et al. Some things happen to cause events just for being there without knowing what they're creating.

Going by the philosophy of causes in nature, so far, four natural agents, factors or conditions have been identified as causing time; it does not really matter much how they're described. The reality is that beginning from atoms things are caused by combinations of atoms, whether we as human beings do experience them or not, particularly quantum mechanical causes that are buried deep in objects which were never suspected until the dawn of the Einstein Age and QED.

Many people believe that the Einstein Age began with the theories of relativity. In fact, his second greatest discovery was the causes of the photoelectric effect; the first was his theory of time, particularly the discovery that the universe is fragmented and exists by means of different parameters so one system of time cannot be applicable to all of them. Wrongly the general theory of relativity is regarded as his greatest just because of the gravity aspect of it. One can never understand why human beings are so concerned about what happens in interstellar space---it's so far away and not controlled under anybody's conscious direction at all. The British in particular (always sulking and moaning, cursing everybody for the birth of Einstein to come and dethrone Newton), make so much noise about the new theory of gravity.

Ordinarily people think insights about the cosmos are more valuable than insights about the little pieces of matter that actually make-up the mass of our bodies. But to

What is time?

discover that the world of visible matter is actually controlled by the movements of the invisible quanta, is my number one discovery of all time, (the Nobel Committee also agreed), followed by the idea that time itself is created, in the words of Russell, "constructed" by man not imposed by God, and bound to vary from place to place. Gravity comes third in my estimation.

It may sound like fantasy (even treachery to some philosophers), but I think the religions believe the problem of time can never be solved, and that after Darwin time is what makes them believe still that God exists and that he created us. It's no use citing biophysics and all the rest of it. They say time and life go together. We know life because we are living it. We can't know the nature of time though we are using it. It's therefore open to a variety of interpretations. Choosing the logical or scientific interpretations does not preclude others from opting for the religious view. We have to remember that when we use physical cycles, like the yearly cycle, to track time, what we are doing in fact is showing or recording the rate of the passage of mere physical cycles and never the time itself----whatever it may be.

The reason, of course, is that the physical cycles are not time; and we don't even know how long the year is in terms of duration in the mind. They give us just how many times (orbits) the earth has recorded. We count these cycles and call them 'years' or whatever. So that for mankind, ten orbits of the sun is ten years; and one would have deteriorating physique to show for it. Is that time?

Certainly not. We count them as the rate of time but they are mere physical cycles; so it means we count the physical cycles as the rate of the passage of ten years. There is no way that these physical cycles could be the real nature of time, only how it is passing by. The division of the year into months and so forth means nothing---we still cannot define one unit of time. For instance, how long is one second in logic? One minute? One hour? They don't have meaning because we can never know what duration means; counting the years is just a matter of counting cycles, tapping the finger is the same thing. It proves nothing. Perhaps we age as the passage of time. Ah, ten years will make everybody age in some way---more or less. So can we regard ageing as the basis of time? How much do we age as an indication of the passage of one year? There is no clear landmark, and also it varies from person to person.

In fact, ageing is chemical and we actually base our time calculations on ageing without realizing it; time is based on ageing not the other way round. What influences us to call chemical processes time is the regular day and night system, the passage of which we regard as synonymous with the passage of time and the ageing of our bodies, namely, due to religious beliefs we (mistakenly) suppose that the days are passing by due to the natural passage of time in one direction towards the Day of Judgment etc., and as they do so we age 'over time'. This is the universal human belief but completely wrong in both chemistry (meaning science) and logic; it makes the life of those writing logical interpretations of time extremely

What is time?

difficult. For a start, there are no 'regular' days in astronomy or nature at all; there is only one constant day. The rotations of the earth are completely irrelevant---something like human speech or footsteps, they do not count in the interpretation of phenomena because they are not metaphysical entities; like the grunts of cows, or tears of men and women, they do not count. So it means our notions of the passage of time and therefore all traditional ideas about time are wrongly conceived. Every statement that includes the phrase 'turning back time' should be changed. We do not turn time back; it is the chemical process we turn back.

One can see that the religions are no fools. I believe totally in secular time; but those who do not cannot be dismissed as easily as that. And since the religions have money and power they want to keep their privileges with as much mysticism and doubts about nature as possible. That's what they are doing. It's not fair, but there you are. Many things in life are unfair to the losers; but you can't blame wealthy and powerful people for trying to preserve their advantages in life. Life is so nasty that anybody who has any advantages to help him or her live comfortably should seek to preserve it. After all, self preservation is the brain's most powerful, oppressive force on people. You can't avoid it and live.

In any case, here are the purely secular factors or conditions (strictly speaking they're confidence tricks.) We use them for reckoning time for general use either individually or in various combinations. Obviously time is a construction; as such we use the parameters we assign to nature or recognize in our environments to construct it, and therefore bound to be different on other planets. But time anywhere else in the universe will have been 'constructed' the same way—by the uses of either one or many of the following agents. The Russellian notion that time is a construction seems the most logical explanation of time under relativity. Not that time cannot exist under the Einstein proposal---that was an expression of intellectual laziness---but that it is constructed by the human brain. So it's the brain we have to decipher atom by atom, if we can! Not only about time, why is it able to probe the entire cosmos? The whole idea of being in existence, of being human, centres on the brain's extraordinary incisive powers. What is the purpose anyway, since we are so ephemeral and infinitesimal? For me it shows the infinite generation and regeneration, death and rebirth round and round indefinitely. Pythagoras was nearly right, except that the same material does not last; atoms disintegrate. Otherwise what is the meaning of regeneration? So, for me, death is sadly the end. Traced from the nature of time, this is philosophy at its deepest level. That's how important I find time to be---even then we can only know how it's passing by and never what it really is. Perhaps the greatest mystery in all nature is time, and this is how it is caused from my point of view:-.

a) Chemistry (physical and organic);

b) Quantum mechanics;

c) Motion;

d) Inertia.

I will now give brief comments about how these agents can cause what we experience as time or a period of waiting, always by means of the brain, the real mystery in all nature.

First, chemistry. The most obvious example of chemistry or chemical processes causing time as a period of waiting is human gestation or any gestation at all; though human pregnancy is what concerns us most. The normal period is nine months. However nine months constitute the time for the growth of a fertilized egg (the fetus) to the point of birth or a normal baby---ready to face the world, with what luck nobody can tell! We call the waiting period 'time' but in chemistry it is a long process of growth, the conversion of matter from one form to another.

Next, the quantum debate. Quantum mechanics are not perceptible, but they cause a period of waiting, though most likely to be very short---it is still time.

The third factor is motion. Motion, obviously, can be seen as time. It takes time to move from A to B. For instance when a sportsman throws a ball into the field of play, it takes time to travel to reach the players in the field. Motion of any kind can show time going; but it is never the real time in force; real time, as I have argued, is never known; only its rate of passage is recorded by the clock together with its psychological effects as the sense of duration. In other words, if real time is unknown, then the cycles we use to reckon time constitute the time in so far as we are concerned. This is what is taking mankind so long to grasp since the abolition of universal time, so much so that nobody will even look at what I write! One agent who did write back said: "I am sorry but not being a professional mathematician or philosopher... I am afraid I cannot figure out why this is important or what it's about..." Brilliant! I am also afraid that it is important because scientific progress (chemistry, medicine, physics, electricity and all that we need for the proper and safe control of our environment) depend on the scientific understanding of time. The problem is that he holds the money, the power and the publicity levers and he does not understand me, or the fact that we need to seek the real nature of things so as to be able to control or manipulate them to our advantage. Even Einstein admitted that he was able to complete the special theory of relativity a few weeks after he got his insight that the Lorentz t_1 is 'time, pure and simply'. Lorentz also said he thought he was unable to discover special relativity because he failed to take his discovery about time as important---and dare I mention that the practical benefits of relativity are too numerous to mention? Without time we can never tract dangerous asteroids.

To go back to our story about the clock, the units produced by the clock are obtained elsewhere and programmed into it for reproduction. The ticking of any clock is not original; it's meticulously conditioned to do it precisely in the form that is done. The units of time are deliberately created (and mathematically divided) to accord

What is time?

with a full orbit of the sun, since we then have to start the orbit of another year. Often people equate motion to the rate of time. Such erroneous statements about the metaphysics of time are all over the place. Every time you mention time you're making a metaphysical statement about the world. It may seem familiar but you could not form such ideas if you're ignorant of the planet's motions. And we do it all the time. Even as I write a newspaper report (The Times, 16/9/13) is repeating the mistake. They wrote: "Flies and small children may have something in common: the ability to slow down time…by seeing the world in slow motion…" This equates motion to time but that is totally wrong. Time, of course, can be based on motion; and every motion can be seen as 'time going'. The problem is how much time? For this reason not every motion is metaphysical time. It is some time going but how much in real time? That can only come from the true nature of time, and that is obtained from the breakdown of the yearly cycle before being programmed into the clock for reproduction in specific units to accord with a full orbit of the sun. It's only when you understand this will you come anywhere near the metaphysical nature of time.

Otherwise if any motion is time then the faster the motion the faster the time and vice versa. That could not give us a stable environment; for all activities are controlled by time. This is additional evidence that motion (any motion) is not time per se, and all statements to the effect that time is mere motion should be regarded as mistaken. For they go on to claim that gravity 'slows' time or speeds it up. We can only do that by affecting the motions of the earth. Real time, again, is unknown. What we do is use mere cyclical motion to show how much time is passing. We are dealing with something almost like quantum mechanics: we cannot demonstrate the entire range of quantum action within matter or atoms that result in chemical changes at the visual level. Yet through the theory of QED we know that they occur and that life depends on them. We couldn't exist without them. Time, of course, is slightly different because we've lived without mechanized time before----when we're very, very primitive---and could probably do so again. (In fact, the historical study of time fails to reveal why it seems so mysterious, because it evolved and we can trace its evolution to its very beginning. But then there is religion, fear of death and the yearning to come back after death together with wild expectations of time travel. Why the human imagination latched on so fiercely to time travel backward and forward beats my understanding. Now they're even saying the Minkowski theory of 4-D geometry makes time travel 'a scientific possibility' through 'curved space-time'.)

I would combine the effects of motion and inertia into one causative agent. Motion and inertia combine to cause activities in stellar bodies. If a planet is moving to or being sucked into a black hole, it could take centuries, or at the very least long enough time for us to get married, raise children, grow old and die; even for several generations to do the same thing; or fight wars; conquer other nations; establish civilizations, lose them, and start all over again!

What is time?

All of these either singly or in combinations can cause time interpreted as a period of waiting---not as 'time allowed' by 'The Creator'. It is an understandable mistake caused by philosophical ignorance to claim that the periods caused by all or any of these alone or in combination is not time like the time assumed to have been bestowed by the mythical Creator of the discredited religions---which means all of them.

Here is my philosophy as regards the act of worshipping deities. The prescriptions (liturgy, Gospels, prayers, incantations, etc.) are revealed by ordinary people. They call them revelations. I call them dreams. In life you have to get education to survive, even as a child you have to be taught how to use the toilet to survive. So education is crucial. But the basic thing we learn in education is how to reason; how, for example, to know that certain objects that resemble food should not be put in the mouth---how to form ideas about things. First we identify things, observe them, establish what they really are as far as possible (e.g. dogs are different from tigers, so that we would not go and play with tigers), and act accordingly, or according to tradition and create safety to yourself and others. This process is called logical reasoning. Without it we would be fantasizing about the world and the things in it, but won't last long because there are so many dangers from snakes, tigers, people and inanimate objects that can very easily put an immediate end to life. Given this precarious conditions of human life on earth (where we are forced to live without knowing why or how we got here), it is unwise to rely on somebody's mere dreams as the prescription to regulate the course of human life, especially en-mass. We know it is unwise because it has led to numerous accidents and ruin. Science gradually evolved after centuries of trial and error to help us identify the nature of things; it teaches us how to deal with them in safety. No religion can do that and therefore all the religions are dangerous---but there is a fear in people that unless they hear something sweet about life and after life they could not live, as they're taught that life is at the mercy of God. Unfortunately delightful religious sermons are not associated with safety and progress; it could even be the reverse, as some people dream that the Gods require human sacrifice and come to take you away. We have learnt through bitter experience that some people are basically evil and infiltrate the religions just to wreak havoc with their fellow human beings' lives; no matter that they utter agreeable sermons---it is a deliberate ploy to seduce people and ruin their lives. For this reason worship of the Gods dreamt up by others is unwise and very, very dangerous. But what about our own dreams, then? Again, dreams are outside logic and therefore unreliable; it is better to stick to what has been rationally examined before you trust your life to it. The unexamined life, we're told, is not worth living.

We are talking here about secular time: after Lorentz and Einstein found that absolute, general or universal time does not exist and running all through the cosmos and the same everywhere, from the past to the present and the future, it became immediately necessary to investigate how our own time began; and the line of thought given above has proved irresistible. It is logically the most convincing reason for the existence of time. It should be emphasized strongly that time was not only thought to be fixed but that it was the same everywhere. That is the reason we think Einstein, being so clever, had a brain wave that if it is changeable then it could not be the same everywhere and so on. This may sound elementary, even tautological, but every little thing about time takes a genius to clarify. Nothing is more mysterious or linguistically intractable.

The Order of Time Seen As a Matter of Arithmetic

Why is time so oppressive that even school children know that time waits for nobody? If they have to go to school at ten, ten it will be, and nobody could delay the time itself---they could delay themselves but not the time. This is the main reason which makes time seem to be an imposition by the Almighty. The religions are very fond of it. We even have a saying in all languages throughout the world that time waits for nobody. The reason, I have found, is this: once we have divided the orbit of the sun into units of time running strictly to coincide with a completed orbit (where 31,536,000 seconds=one year), the time units became oppressive—no one could change them without skewing the alignment with the orbit of the earth round the sun and get into danger, especially during the night in ancient times. That is the reason time follows a strict order, for the units have been carefully worked out to accord with the motions of the earth. The basic cause is the order of arithmetic. Time is no longer as mysterious as it used to be barely a hundred years ago. The oppression and passage of time added to its ordered nature and sequences were the three insoluble aspects of time, making it seem divine; but all these can now be explained logically and conclusively, if we begin from the premise that it is constructed (as Russell deduced).

The human mind craves order and cause, so we think of the order and direction of time—but by who, or imposed by whom? We've just been liberated and freed from the bondage of time's absolutism only to be confronted with its restrictive order and direction. I think it's all the fault of the brain. The brain requires order and direction due to the way it's put together. We imagine that several compounds cobbled together accidentally to create the brain. With each component searching for 'complement', the probing tendency in the brain was established as part of its basic structure. This may be mere speculation, yes; but that is how we've gained all the knowledge we possess.

Even then, there is some tenuous evidence for this because we can see how it grows in the fetus as the host (the woman's womb) supplied it with the necessary compounds and chemistry, and when complete, begins to take over the fetus and cause its growth to the point of birth; after which it is fed externally to direct the body to the point of death. Originally it must have grown compound by compound. The process probably was that one

What is time?

compound and another joined up accidentally. Then the quest for 'complements', the need for order and the sense of waiting resulted, caused by what I can only describe as 'chemical hunger'. Most of what is written here cannot be proved; but I think it may come close to what actually happened. For something must have happened to cause the generation of the brain out of inanimate matter, elements or compounds. This is an attempt by ordinary human beings to trace the physical origins of the human brain. It's not rocket science, as they say. The important thing is that it's not religion or fantasy either, for the life itself grew the same way: elements formed the original, egg or sperm to create the basic primeval amoeba that replicated till it grew to be a sustainable organism. It is also evident that the brain's demands on the human body (including forcing us to have and endure the sense of waiting which we know as time), are built into the brain's structure; and we imagine that it could have come from the protracted, chemical processing over many centuries involved in the brain's creation out of the elements. This word 'element' is used instead of saying atoms, but it is to be understood that its ultimate constituents are believed to be atoms. This, of course, is the language of science. It is conceded that not everybody believes in science; what is evident, however, is that nobody can live without science. Some of us abide by it on pain of death. Otherwise it's not illegal to believe in anything that causes no harm to others.

To continue with our story of how the brain probably evolved, we can imagine that when the required compound for complement arrived, more chemical tentacles were created---for the brain is more than a million times more complex than the biggest computer ever created by man---more materials were needed for a complete organism to reach an independent and sustainable whole. Hunger, the sense of waiting, yearning, cravings, the need for reason to explain causes and the need for completion to calm the increasing number of chemical tentacles requiring 'soothing' (roughly, it is thought), caused the brain to come to exist with these tendencies built into it. One of them is time, or the sense of waiting (waiting for the require complements), which internally we know as duration. Another is the craving for order and causes. They do not exist in the universe outside the human mind, that is the reason the cosmos seems so chaotic---I believe it's not governed by time or order but by accident and chance. Even the world is similarly chaotic, unless a human being is involved or is controlling events.

Otherwise by whose order or direction is the putative arrow of time following? It may very well be that the order and direction of time are misleading concepts, precisely the manner we get the years and the centuries. Time does not seem to move physically only mathematically; you can easily alter ten year to twenty on paper or through mathematics; but the physical aspects of time are external and come from the motions of the earth. Thus altering ages (or the figures of time) on paper is useless; the physics will always win by skewing the human estimates. After that human ingenuity took over to create concepts of time in

What is time?

the mind to accord with the motions of the earth, which may be regarded as the birth or the cause of the birth of our whole concept of time. Since the earth's motions are repetitive, the units of time spawned by them are also digital, and we applied arithmetic to them. Thus, unlike our primitive ancestors' practice of keeping time with charcoal marks on the wall to indicate the number of days, weeks, and months gone by, as the passage of time, modern man has the use of theory to simplify things for him---after that the mystery-makers took over, beginning with Pythagoras.

Anyway, because of all this, we now know that time is produced in units and the units multiply for it to advance or move on. Is that still disputed, with the example of the yearly cycles being so clearly demonstrated? But it's essentially a matter of arithmetic: if the year is one unit of time then it replicates to become two, there, four, five, six and so forth all the way to the centuries. The yearly cycle is not moving past the signposts of the years all the way to the centuries; this is obviously not the case. Rather the year is repeated over and over again indefinitely. Is that still disputed by the mystery (or mischief) makers in the universities? It has the merit of confirming Russell's idea that we actually do construct our own time, probably through the brain's instinctive cravings for order; so the order of time is in the mind not in the world out there. Sentience, a theory of numbers, the ability to count and points are required to fulfill this order. Hence space is involved, since the use of points implies space for the creation of theoretical time based on mathematics or arithmetic. We can still call it 'space-time' but only in the sense that time is or can only be created in association with space; and its passage, too, is the same as its ordered progression---i.e. through the procession of its units, a matter of arithmetic. The years increase in numbers to pass by, without any direction. For one thing we've been able to establish clearly is that things are created continually in the universe through the accidental combinations of elements, compounds, atoms, et al. Even the brain created itself and dies to prove that it is not a permanent entity, just a passing thing produced through accidental causes.

All the same, some writers have made the order of time the pivot of their own interpretation of time, even though the definition of time as the 'irreversible general passage of existence' was meant to refer to a time system imposed on us to run generally in one direction through the entire universe and the same everywhere, with the condition that we are all moving in tandem to the Day of Judgment as the end of time---the biggest folly in the human mind's suppositions, incredibly, illogical and totally without foundation, just part of the Christians' meaningless fables that we have wisely rejected for over a century ago.

After its rejection the order of time should have been seen as a purely logical matter easily resolved with arithmetic. Let me explain. The phrase and particularly the word 'irreversible' mislead people into thinking that we are all moving irreversibly with time (plants, animals, entropy, vegetation, the seas, rivers and streams, et al) to a predetermined end or destination. This has spawned numerous legends, theories and beliefs mistakenly; yet the order of time merely means the arithmetical order or progression of time's units---the years, for instance; exactly like counting the years from one to a century. And one year is also pared down to the seconds, that should reach a certain number (counted progressively in arithmetic), to coincide with a complete orbit of the sun---and start again. Because of the restart, the number of units counted has to be exactly correct; therefore time cannot wait for anybody. It really cannot do so physically. The phrase does not mean an irreversible passage of time that we cannot interfere with, but irreversibly leading us to doom. This is what scientists have magnified into the theory of entropy's irresistible march to the death of all activity. The theory of time has accumulated thousands of myths, fables, fantasies, legends, lies, religious beliefs and even mischief. I am quite sure several sacrifices have been carried out with human beings about time to placate the Gods. Yet the order of time which some writers regard as an insoluble problem implying divine influence is nothing more than the progressive counting of the units of time, after all that is how we get the centuries.

The order of time and the supposed irreversible passage of time should have been eliminated from the debate as soon as we realized that time is discrete as intervals or time units between points; for obviously we do not all move as such: the Heaven is nowhere; God is supposed to be dead; and existence is not even uniform, neither do we all move in one direction. In quantum theory the direction of motion is not even known. Some things are stationary, others are moving in reverse, and others are moving in any way they prefer. Even in the solar system presumably controlled by the sun's gravitational attractions, not all the planets move in one direction. In any case, once we found that time is variable, it was unwise to insist that we are under the command of one kind of time moving in one kind of direction to a solitary mythical destination. We now know that time, existence and motion are all variable. This is the situation Einstein discovered, namely physics is for the planets because there is only chaos among the stars (too big and complex for anything else). Time also belongs to the planets. For the cosmos to have regulated time somebody must live among the stars who has the brains to apprehend cyclical motions that could be used to reckon time sequences.

In spite of all this, everybody comes to the study of time with his or her own agenda without reference to logical truth because religion and ancient traditions have so conditioned our minds that we all think we know what time is. Einstein alone showed (he did not just say it; he proved it by experiments) that we are all wrong, and Bertrand Russell not only agreed with him but said that his theory of time was, perhaps, his greatest achievement. For me there is no doubt about it. General relativity is not Einstein's greatest achievement. It's not even his second. It's his third after time and the quantum theory. I am of the opinion that what we find in interstellar space is of secondary importance to what we find here on earth; for even if we

What is time?

discover a body on a collision course it is what we find on earth that could be used to neutralize it. In most cases interstellar knowledge is mere intellectual pastime for selfish, psychological satisfaction. Something much more like an ordinary labour of love. The near star mentioned above may explode and destroy not only life on earth but the earth itself. But what can cosmologist do about it? --- absolutely nothing; as I have said, except to give us sufficient warning to go and join Richard Branson in his special plane, but even then where to? Could there be any permanently safe place in the cosmos?---to want to live forever is certainly not a rational idea, We are born to die, but why? What for?

I know that human vanity is bigger than the sky, but obviously the universe is just too big and complex for us to worry about putting anything right up there. The causes of its nature and activities can hardly be more important than the price of bread here on earth. No matter what we do or believe no course of action by human beings (as insignificant as we are) can make any difference (other than, perhaps, diverting or destroying asteroids on a collision course.)

Although not a believer, I sympathize (somewhat) with believers about the purpose of human existence---what for? Man is so insignificant. Even the planet itself is just a tiny dot soon to end up in a black hole and burn out of existence; yet human intelligence is so far-reaching, so inquisitive, so hungry for knowledge of the cosmos (probing, probing, and probing), to no avail. For all his insight about the cosmos, Einstein died, decomposed and disappeared out of existence altogether. But his insights and discoveries about the world we live in are in daily use for the benefit of mankind. They certainly are more important than knowing about black holes. So while the human brain's creations can last, we the creators, the bearers of the brains, could die easily. Men are so fragile. The religious people are not that stupid---there is a real problem with human life on earth; there is human yearning for salvation; we just do not want to come into the world to die in misery---what is the point of that? The religions want to claw at the tiniest straw that could be interpreted as giving meaning to the senseless thing called life: the best philosophy for the worst reasons. But the alternative, as we know, is sheer emptiness, misery and early death for no sensible purpose whatsoever. Thus, despite my basic irreligious beliefs, I still think religious tolerance is the beginning of genuine humanitarian wisdom---"Remember your humanity and forget the rest", was the last advice Bertrand Russell gave us. And Russell, for the ignorant, was not just another human being. He was the world's greatest philosopher at the time, and most likely as clever as Aristotle.

However, as the result of the new Einstein theory of time, we now know that there is truly no longer a universal time as Bertrand Russell put it in his book ABC of Relativity; so we have got to search for the mechanism of time or how we get our time--- or what we call time. Atomic time is included in earth time, as part of the time we have created with the earth's motions. It is often assumed by some religious scientists that atomic time constitutes a cast-iron proof that time exists in the cosmos and can be measured in many different ways. In fact, atomic time is not different from earth time. The cycles, pulses or oscillations are merely shorter than the long orbit of the sun, and, in any case, they have always to be related to the second to make sense. For this reason atomic time is still part of earth time. One can even tap the finger. It is the same thing---something we can count as the rate of the passage of time is all we can have for the reckoning of time. It is through sheer hard work and amazing human ingenuity (mostly by the mathematicians), that we have 'constructed' what we call time to guide our activities; and since this time is based on the earth's conditions, not all of which are conducive (or compatible) to living without sensible controls, our time is strictly tuned to show us the safe periods and areas of the world's conditions and environments we can negotiate in safety.

Otherwise there is no time in the cosmos at large. Our time is unique; there is no doubt about it. For it is created with the unique parameters of the earth. If there is natural time behind the parameters we simply cannot know it, because that is not what we know as 'our time'. What we call time, or our time, is 'constructed' from the parameters as their effects only---counting physical cycles as 'years' is not time. Perhaps they are the effects caused by natural time behind the parameters, but we simply do not know.

As already mentioned, Einstein divided the universe into two. They are the metric of general relativity where there is no tolerable conditions for human lives, and the metric of special relativity where life is feasible. In this home of ours there can be time, as we suppose that in similar homes elsewhere in the cosmos time will be thinkable. Where there is life, there will be time. That is part of the logic of time in the universe. Every time system can only be based on physical parameters; and they are all different one from another. Some or most of these parameters are present in all the segments of the universe; otherwise there is no time in the universe. To have time you've got to have the intelligence to construct one out of the components of the relevant parameters.

The conundrum is this: on the one hand, we think there is something called time naturally moving on by means of the factors or agents mentioned above; but if so, what then is moving these agent on? On the other, it would appear that, like the brain emerging from nowhere and seizing control of everything in sight till its own demise, human ingenuity has created what we call time out of the natural features (or parameters) we find in the universe. It appears these parameters or features, being mere physical materials, would know nothing about time as the sense of duration in our minds.

Being the greatest philosopher of the period under discussion, Russell asked the most important question about time, enough to redeem philosophers' reputation after their condemnation by Karl Popper. When Lorentz

What is time?

and Einstein showed that absolute, fixed or general time permeating the whole cosmos (and the same everywhere) does not exist, Russell asked, what then is measured by the clock? Frankly, apart from the orbits of the sun, there is nothing (unless you can tap your fingers continually as the rates of the passage of time). Hence the thought of secular time.

A careful analysis of the Russell question gives a perfectly logical explanation of all aspects of time as a secular entity 'constructed' for use on this inertial frame, and even then only capable of showing how much time is passing and never what it is---provided one can ignore the billion or so myths about time. We use cyclical or regular motions to give us time--but they are physical, so they can only show how much physical cycles are passing (have passed or will come to pass, e.g. as 'years'), and we use them as time periods to plan all activities: ten hours means it is time to do so-and-so for so much hours, etc. What the real time 'is' we can never find out, only how many cycles (being the years) of it have passed or are passing. As I have said, my guess is that time is a combination of chemistry and motion (especially repetitive motions) and sentience; none of these on its own can be mechanized into a clock as time, but in combination they can give one 'a period of waiting' (especially in chemistry), which is time. Sentience is required because somebody must be there to set the points and count the orbits of the sun as years or there will be no years and seconds derived as fractions of the year. Until we were wise enough to do so, man had no time and lived like a beast of the forests. Just look at the story of the evolution of the clock since we came down from the trees.

Of late many books have come on the market dealing with time travel, and it seems they have been very successful, so successful that publishers fail to notice the contradictions in their theories. I am ashamed by this trend in publishing (where even convicted murderers are given millions to tell their stories).

For instance, in A World Without Time (Penguin, 2007, already cited) Professor Palle Yourgrau states categorically that Kurt Gödel has "proved that in any universe ruled by the Theory of Relativity, time simply cannot exist..." At another page he says it has been proved that time travel is "a scientific possibility", and continues, like the rest of us, to live in Einstein's Special Relativity frame where he says "time simply cannot exist". It is difficult to see the logic in saying anybody who is not a magician can travel by something that does not exist. It's appropriate that the book is called A World Without Time. To me it's a world of fantasy, for this world we live in certainly has time: all workers go to work by time.

One implication of all these contradictions is that the phrase 'Space-time' may be sensible, succinct and cute to some scientists and those mathematicians who want to reject the 3+1 formula for representing physical reality in space, but it's not actually true of the physical world, unless it means time can only be gained through the application of points to space, so that we get time units (or time intervals) as relation between points, like the years, not in the sense

What is time?

that space and time constitute one entity---just to avoid use of the 3+1 formula--- so that as space curved in general relativity it would take time with it, as 'curved space-time' for you to meet your grandparents even before they were married, just to justify religious sermons about time travel being 'a scientific possibility'.

It is obvious that the universe has no time as an oppressively unavoidable order of action, as we have on earth that is why we get problems with the quantum and other sub-atomic particles. Neils Bohr said whoever is not shocked by the quantum theory has not understood it. That statement should be turned on its head, namely whoever is shocked by the quantum's strange behavior has not understood it---he or she does not realize that the quantum alone (without conscious direction as in LASER) is not subject to time or what we call human time, 'constructed', out of matter after the quantum came to be in existence. I would advice that we look at the quantum carefully. From what we know of its nature, we understand that it would have been there (in a strange sort of existence we can never imagine) long before their interactions caused objects to come to be in existence. It can be in two places at the same time because it is the most natural piece of matter behaving without the influence of time; it is outside order in nature, it is not directed by anything; the human mind's notions of order and time sequences do not apply to it. You would be shocked by its strange behavior because you can only judge it with a shallow mind that came to exist after the quantum and is therefore unknown or recognized by it. This is looking deep (speculatively) into matter to the quantum level, so deep that physics cannot include it---but that is how physics itself came to exist and yet it works to the extent of having the capacity to destroy all life on earth.

As the most original matter and the smallest bit of matter that can exist, the quantum is not subject to any of the human concepts about order, time, motion and chemistry as we know them. The quantum existed before the regular cycles we use for time, order, motion, chemistry. This is how Feynman put the same idea: "The word 'quantum' refers to this peculiar aspect of nature that goes against common sense"---exactly. It belongs to a universe before the common sense came to exist and is therefore not subject to any of its notions. Common sense refers to common objects of the perceptible universe. The quantum is not part of this universe. It's the most original and basic matter whose many and varied interactions have created phenomena as we learn from QED; it therefore does not know how to behave to suit us as we are part of that phenomena to which it does not belong.

We can construct time out of the phenomena in our experience, that's our peculiar luck or curse. It all depends on how you look at it. All the elements for this act of construction are there everywhere in the universe but not as time (to the universe). They are rather events occurring haphazardly under a variety of forces: gravity, space, inertia, motion, heat, chemistry, without conscious control. The religions are right about one thing: the process of human creation requires intelligence. Where did it come

What is time?

from? They claim to know that, but cannot prove how they know it. References to the scriptures are what annoy scientists most.

Again, our instincts expect the quantum to respect time or behave according to time sequences---but the universe has no time. The parameters we use to reckon time are purely accidental events. We know they occur, but cannot think of how anybody could have organized them in the manner we arrange things on earth. As we have come to realize, there seems to be no direction in the universe, no purpose and no logical sequences. Existence is existence; it's just there. Life came as a chemical accident, but, like everything else, it just happened for no purpose at all; and while it seems to us to have been long in existence, in the cosmos at large our period of existence is just a flash, and the earth itself just an infinitesimal dot, not worth bothering about. We see nobody there to worry about it either, unless and until we set things and events in earth-time. Ah, but the universe has no time! Earth time is just that, namely a time system we have created for ourselves on this planet and applicable to this planet alone. That's the conundrum, and I for one find it enormously interesting just pondering it, usually alone, as my religion. Some people weep over life's problems. My advice is to try and find them interesting as events occurring without cosmic control or significance, and yet so vast and complex that pondering them is itself rewarding as an intoxicating spiritual solace. It's not true that it will make you mad---it'll rather cure your madness!

Nobody is there to infect you with the germs of madness. In nature things happen haphazardly. There are some limited logical sequences like something cold fleeing something hot or ice melting at certain temperatures, but no streams of logical sequences (such as we human beings can construct out of this huge and complex admixture of accidental events), by means of the human mind including the consideration of time---the most essential thing besides life. So it appears that outside a human head time does not exist. According to Russell we 'construct' it ourselves. I agree with Russell absolutely. I do not believe that time does not exist on earth because we are using time everyday; but it certainly cannot be defined logically. For instance, how long is one year defined as a unit of time? When one year passes you know that you've aged one more year but how long is that? Try as we may, we can never define the year on its own logically without using any of its fractions (say, the months or days, which is logically unacceptable).

It's obvious that time does exist on earth; the problem is how to define it. So serious is this problem that we have come to the conclusion that its logical definition amounts to just how it is passing by. So we think we can only know how it is passing by and never what it is. But in the universe at large, although we can spot some of the parameters we use for time on earth (everywhere), no one is constructing time sequences out there through the use of these elements, not from our point of view anyway. All notions of

What is time?

time are carried from earth to apply to the cosmos in breach of the Einstein theory of frames.

Chapter Five: Conditioning The Human Mind For Time

I regard this as the appropriate juncture to explain the nature and importance of time before we carry on. I agree there are past, present and future in the world or the universe, of course. Einstein probably meant they're human concepts for our convenience, not basic and permanent episodes to be allowed to influence theory; that's my understanding of what he said: the past is obviously what has occurred most of which, but not all, would exist only in memory; the present is what is on-going, and the future is

What is time?

what we would expect to be the consequences of the present and past put together. Logically all this is beyond dispute. However, as related to the interpretation of phenomena they do not count; they are not relevant at all, and can only lead to confusion and wrong ideas about things. For they are based on the concept of 'passing days and nights', which constitute man's notion of the passage of time. Yet there are no passing days and nights in actual, physical reality in the whole of the universe. There is only one constant day anywhere, and it never moves, dims nor closes down even for a second. If it does life will be extinguished. The days and nights are temporary blips caused by the earth's revolutions but they do not count because the earth is so insignificant. There are so many billions of gigantic bodies (stars, suns, moons, etc.), that if we were to use any one body's motions to interpret phenomena, we could end up changing our ideas so regularly that stable life would not be attainable. What the earth leads or misleads us to suppose cannot be used to interpret or influence the cosmos. Above all nothing is passing through the universe in some kind of a thread, least of all time, which consists of separate units (like the years) and passes by through the succession of its units, again, like the years. Hence all other units derived as fractions of the year are also separate and individual and succeed one another. This is the most logical explanation of time we have for scientific use, deduced from just four thinkers in all history, namely Einstein, Leibniz, Bertrand Russell and Professor A.N. Whitehead.

But still speculating about time, I think if it is true that every inertial body has to have its own time, then there simply is no time in any part of the universe until you have created your own time and conditioned your mind to its nature, and Einstein made it clear that we can only do so in inertial bodies, not in general relativity. Conditioning our minds with our time is bound to lead to changes not naturally existing in other parts of the universe. The elements we need to construct our time are not available in general relativity; and if they are not available even in general relativity then the other world the quantum came from would not have them either. As noted, Einstein divided the cosmos into two: one is where you can have time by sustained regularities (or 'constructed' logical sequences lasting long enough for the human mind to use for its creations), and the other is where, because of the strong gravity, you cannot even see anything anywhere at all to have the necessary regularities of motion to use for time. For time means from when to when, from one point to another---the year, for instance. Part of the problem in physics come from the improper understanding of Einstein's ideas, for all time is based on regular or repetitive cycles---the year, for instance, and since it is the cycles we count as the rate of the passage of time (like the years) we can never know the true nature of time only how it is passing by. Through our mathematical ingenuity, we've learnt to use cycles to provide units of time. I call this the quantification of time. These are what we use as time: years, hours, minutes and so forth. They merely indicate how time is passing by, obviating the need for complicated

What is time?

theories about how time passes through nature. For a start, our time is not even passing by, or passing through nature. It is discrete and proceeds unit by unit---year by year, minute by minute and so on. But in my experience it seems mankind doesn't want to know about discrete time. Man is so enthralled with time passing through nature to the Day of Judgment, and start all over again due to the transmigration of souls. It seems nobody wants to die if he'd not come back to life after his holiday up there! Or live in another world up there. Thus it's easy to get some religious people to commit acts of terrorism in suicidal attacks.

Let me emphasize again that what we call time is only how it is passing by---the years for instance, pared down to the seconds. But Einstein is so misunderstood that many writers insist that we need the Minkowski formula to understand relativity; yet the special theory of relativity that concerns us most on this planet had nothing to do with the general theory of relativity and Einstein's use of the four-dimensional continuum of Minkowski in the equations of general relativity. Rather it merely amounts to incorporating time into space to form one entity so as to dispense with the 3+1 formula. There are suggestions about the usefulness of this procedure, but originally it formed no part of the theory of special relativity. In other words, the Einstein theory was complete without Minkowski in so far as special relativity is concerned. The Minkowski theory did not improve special relativity---it's already complete and critically acclaimed. All suggestions to the contrary is evidence of ignorance; for it is obvious to me that only those working closely with relativity understand it without using the Minkowski formula for equating space to time in one equation, yet it was never successful. Professor Eddington was right when he struggled to recall the name of the putative third professor who understood the theory in the initial stages. Looking back, I believe it was himself, Bertrand Russell and Professor Whitehead. I don't think Minkowski ever recognized that 4-D geometry is no help to relativity, meaning that he therefore did not really understand relativity. The culprit is always religion; even the best thinkers cannot free themselves of the thought that time is eternal and imposed by Providence. Otherwise it is difficult to see how any intelligent man can accept that time is incorporated in a universal entity like geometry. It is an ephemeral entity created for the convenience of ephemeral mankind.

Let us look at time in practical terms. Our parents agonize about what will happen to us when they are dead, because they know life would continue as it had happened to them when their own parents died. So when they're gone, we would be there. It means the world will carry on when we are also gone. If the world carries on it means geometry is carrying on. Yet our time will not carry on when there is no one there to set the points for the cyclical units we call 'one year', and out of which all time is derived as fraction. In plain words, our time will end with the earth's demise. How therefore can this ephemeral time be part of geometry which, of course, is eternal? The theory should

What is time?

stop at deriving time from space with points, meaning the time could not exist without the space---and therefore it is 'space-time'.

Presently as I see it, the religious people want to resurrect the bogey of past, present and future to prove Einstein wrong. They claim that the syndrome causes the flow of the story of history; that history is what the study of the past to the present tells us. Yet past, present and future can be perfectly logically explained as uneducated fiction, so Einstein was right: the past is obviously memory; the present is now, carrying the past as historical baggage with it (you never leave your problems or wealth behind you, do you?); and the future is mere speculation. History is the march of these events not time; and the events are still marching on as the continuing story of life. Thus the past is not still existing anywhere to be revisited---it is here with us as the consequences of what happened in the past! History is not seen as time running through nature from the past to the present and so forth. Our time, as the year shows, is discrete. Discrete time cannot march throughout history (or the cosmos) as people like to believe. Only events have antecedents and consequences. The times are added as the times of occurrence. It is the events that mater. Many of the religious-based mysteries of time can also be resolved. I have published ten monographs about post-relativity time explaining all these issues, but nobody is showing any interest as people continue to chase their emotional thrills from worthless books and gadgets. I fear that true culture is dying slowly due to the aggressive onslaught of the electronic strangulation.

If time is not permeating the cosmos and moving from the past to the present and going on to the future then there is no quandary. You can challenge this theory of time, but by the yearly cycle we should know that time is discrete, from year to year, repeated over and over again for all the centuries; there is only one year in all the universe, repeated to carry on as years; and every unit of time, too, is derived from the year together with its astronomical features. If time is like a thread passing through nature, then it is reasonable to search for theories to account for how it is passing. But we have a time system that is repeated to continue. The year is only one; to have two years we repeat it; to have a thousand we go round the sun a thousand times. Surely everybody can understand how this time passes by as something in procession---units of time following each other? There's no need for a theory to explain how time passes by. This is all we call time. Every unit of time is a fraction of the year as divided with points or astronomical features. Time units have no independent existence; they exist only as fractions of the year no matter how they are derived. The mathematicians have done a good job about this; but there is no mystery; it's plain common sense.

We've nothing else for the reckoning of time except mythologies or counting the days as the passage of time in a primitive manner without theories. I must repeat that there are no years at all in nature existing as something we

What is time?

can just pluck out of the sky and apply to events. Let me repeat again that there is only one year and all other units of time are fractions of the year. To have more years we simply go round the sun again and again; that is what we know as the passage of time, or it constitutes the passage of time---namely the units of time in procession. This idea solves at once the fearful conundrum of the passage of time.

We hear so much from writers about the passage of time, but nobody has ever been able to define time. You have to define something before knowing how it behaves. Once time is defined as relation between points, like the year (and that is not in dispute because that is how we get the year, and the year is time), it becomes something proceeding unit by unit or intervals of time in procession causing the continuity of time---like the year increasing in numbers all the way to the centuries. After all, human notions of time come from the year and daily revolutions of the earth, or the day and night system. This time can only proceed unit by unit; so the passage of time is seen as the procession of time units---the year for instance, worth repeating a thousand times to defeat the stubborn doubters and critics of rational thought. And as these units are all passing it means all we can ever know of time is how it is passing by and theories of the passage of time are redundant, even humbug when it's incorporated in religious sermons and the Day of Judgment mythology.

Even then (strictly speaking) going round the sun is not time. We use it to show how much time is passing and never the true nature of time; for going round the sun is a physical activity, yet we count them as years because we have nothing else to indicate how much time is passing. The years replicate to become centuries; or they increase in numbers to pass by. Using the yearly cycle to know how time is passing is not time that is passing. Time does not pass, only the units do; but the units are mere physical cycles. "A time system", as Professor A.N. Whitehead has said, "is a sequence of non-interacting moments"----year after year after year, or as pared down to the seconds and the other fractions of the year, and he made this observation in his book entitled The Principle of Relativity---I am not afraid of these repetitions because this is a theory about time; a subject so contentious that new ideas about time ought to be hammered home or fail to convince the critics, no matter how true they may actually be. On the other hand if my critics have got the message they should tell me so that I can relax or even retire altogether!

Everybody on earth agrees that time is mysterious, yet the scientific study of time after Einstein is becoming logically consistent and even delectable as logical solutions delight us all; the human mind craves logical thought; and if we reason logically from the premise that there is no longer a universal or cosmic time, and that the basic unit of our time is not even definable, so we simply do not know how long the year is and therefore all measures of cosmic time are flawed, then time becomes easy to understand. We normally say human beings age 'over time', and the year is regarded as the best yardstick of age and ageing. The truth

What is time?

is that the year has nothing to do with ageing. We age through chemistry and metabolism but they take time to occur. Everything takes time; that does not mean time is causing them, the underlying causes are always there if we look hard enough. Time takes the credit or blame out of ignorance simply because it is always there as the chemical, accidental and physical causes of events, these we choose to call 'periods of time', again out of ignorance.

But we can't tell this to the cosmologists wasting the taxpayer's money on their pet projects---like smashing atoms. They say it brings technological, economic and medical spin-offs---yes indeed they do, but researchers could achieve the same results over time without wasting billions. How many billions did Bill Gates spend before hitting on his ideas for lucrative ventures?

Brains are what we need. In my opinion the CERN is a complete waste of money. And they're not even sure of their theory. They announced recently that if c has been breached then they might have to re-examine the concept of 4-D geometry to see if it is really true. The point is that the Minkowski ict equation is flawed for being based on the imaginary time coordinate, i, not because of the status of c.

Being a grumpy and infirm 78-year-old great grandfather, I am too old to fear of what could happen to my career; others are probably silent because they have good reason to fear the powers that be! Scientists chasing research funds are as ruthless as the Mafia, probably more so. But there is a lesson here. Science is different from any other calling. It is so open that any theory in any branch of science that is not true cannot be hidden for long.

In case there is any doubt about the Einstein theory of time which states that "There are as many times as there are inertial frames", simply because the parameters used for 'constructing' time are different from one place to another, let me repeat what Professor Eddington said about the matter as a reminder: "Prior to Einstein's researches no doubt was entertained that there existed a 'true even-flowing time' which was unique and universal...Those who still insist on the existence of a unique 'true time' generally rely on the possibility that the resources of experiment are not yet exhausted and that someday a discriminating test may be found. But the off-chance that a future generation may discover a significance in our utterances is scarcely an excuse for making meaningless noises." (Mathematical Theory of Relativity, Ch.1.1.) I repeat this on purpose.

Scientific mysticism has always been part of the problem of time's definition, but now, thanks to Eddington, everybody can be sure that there is no such monster called 'Time Zero' from whence time is supposed to have began and running all through the universe ever since from the past to the present and the future till God calls a halt to the whole damn thing on the day of judgment; a childish fable forged on us as the true and most profound philosophy of existence. To my mind this is intellectually shameful. Religious believers may be gullible, but the rest of us are not that childish to believe an infantile fable like that.

By the same token, 'curved space-time' by which time travel is said to be 'a scientific possibility' by the very people awarding the Eddington Gold Medal to their clever fellows is totally untrue. But sure, physics has got to put its house in order. And the physicists must start with no illusions about our time having any influence in the universe, for the whole earth is only a tiny, tiny, tiny little infinitesimal dot in the milky way, let alone the entire universe; and the psychological-time constructed by the 'worms of the dust' crawling on its surface (as the poets describe us), can hardly influence the universe at large, although I know that greedy publishers wanting to cash in about time travel based on human gullibility, will continue to publish such books (as mentioned above), implying that we and our time can have some kind of influence on the cosmos as a whole. There ought to be a law against the spread of such falsehoods that make people less not more rational. The universe is certainly mysterious; but time is not so strange any more, since we know that at some stage in our lives we simple had no time because we lived like apes on trees. Since we came down we have tried many things for telling the time, the most rational of which is the earth year, pared down to the seconds and the atomic pulses---that's the most logical explanation of time possible and we owe it to Albert Einstein alone. I will now try to sketch what time was before Einstein and what it became after the great man.

Chapter Six: The Nature Of Time Before And After Einstein

Before Einstein time was supposed to be general and absolute, such that any unit of time here is the same everywhere else, a creation ultimately attributed to God. After Einstein we see it as a secular entity that is limited to a frame and discrete because we can reckon its passing only with repetitive cycles. These cycles can only give us discrete units of time, year after year after year and so forth. We pare the year down to fractions ending in the seconds and the atomic oscillations based on the second.

What is time?

As some writers have observed, Einstein's theory of time arose from experimental results and therefore not open to doubt. Now let us look at time before and after the great man in a little detail.

(a) - TIME BEFORE EINSTEIN

Generally speaking, few writers have studied time seriously with a view to suggesting cogent theories about its nature before Einstein. Even those who made such attempts, like Henry Bergson, were guilty of assuming that it is just there. That we find it in existence, and that is that. Even the careful, logical thinkers fared no better. They made it look synonymous with motion or 'Being', eventually calling it the "irreversible passage of existence". Yet existence is not one; it is multitudinous and individual. This means at any moment billions of movements are taking place: some (like leaves, just waving in the air), moving up and down, sideways, forwards and backwards, tumbling, limping, dancing, rolling, crawling. Moreover every individual is uniquely separate with his or her own perspectives---no two persons, as Einstein showed with his analysis of simultaneity, perceive one event identically---space and time coordinates are involved. The multitudes of people perceive the world differently. Above all, existence is not altogether passing in tandem. In quantum physics directions are not even fixed; what may be irreversible to you could be the opposite to somebody else looking at the save event from another angle. It is necessary to mention that Einstein also failed to decide how time is created, and stressed only that it is neither fixed nor absolute, adding the most revolutionary idea that it originated from this inertial frame and that there could be as many times as there are inertial frames, thus laying the foundation of secular time.

To digress a little here (taking liberties, the reader might say!) to discuss the thinkers I commend, I have to stress that, contrary to the opinions of some scientists and particularly the pure mathematicians, philosophy is not as time- wasting as is generally put about; it is so serious that it shares with theoretical physics the ultimate attempt to formulate credible theories for our understanding of the nature of the external world, or the cosmos as a whole, including psychology and cosmology, or the mind and life ---what happens in our heads and the entire universe of sentient beings. There have been great and valuable contributions to human welfare and material progress from philosophers. Apart from what one may describe as 'the scientific thinkers' like Pascal, Archimedes, Aristotle, Pythagoras and the rest, only a few philosophers have actually made valuable contributions to human progress, but they are there. I am thinking of writers like Plato (with all his faults he discovered the idea that we can never perceive reality as it really is by his most original, first-class supposition called 'The Simile of the Cave'. He also promoted philosophy and taught Aristotle.) Rene Descartes' merits included the Coordinate Geometry, upon which relativity is built, and of course, The Cogito. Our own

What is time?

Bertrand Russell's many merits include the new logical theory of 'Denoting'. Henri Bergson speculated that space and time are separate entities. If we had listened to him we would not have wasted so much time and energy over the Minkowski fiction that they are one entity---which, I believe, is still causing distortions in physics and relativity. Professor A.N. Whitehead and Bertrand Russell also discovered that the world of sense is 'a construction not an inference', which liberated scientific thought to an enormous degree and still bearing fruit. For example, the photons 'construct' images; this idea can be seen as much more rational than supposing that we perceive pre-existing images created by God, which we are only able to see by the grace of God as they are invoked by the mind out of thin air as Plato proposed in his theory of Ideas, and which has been used to justify the existence of God by the religions. In monumental efforts, Russell and Whitehead again tried to derive all mathematics and arithmetic from logical premises, and even though it's judge to be unsuccessful, many logicians believe that it inspired Kurt Gödel's Incompleteness Theorem, otherwise known as The Theorem or The Proof. However, as we have seen, Gödel later ruined his contribution with suggestions that time travel (a notorious religious myth) may be 'scientifically possible' and that even Einstein agreed with him!

The Cartesian Dualism and his notion of God as the Absolute Perfect Being are not mentioned because they are religious notions and religion causes wars and does not advance philosophical knowledge. All the religions borrow from two basic ideas. Due to the fear of death as the end of human life, they borrow from the Pythagorean 'Transmigration of Souls'; and also from the Rene Descartes dichotomy between the soul and the body of man, or Cartesian Dualism as it's called. We don't want any of that because it promotes religion and religion causes wars. Human life is bedevil by a fantastic paradox: the majority of mankind believe that they need to worship and couldn't live without it, quite apart from the hope that it could make death just a transition to a better life; yet the organizations they establish for this worship cause wars and sectarian conflicts that make peaceful life on this planet so hazardous that even if life after death were possible it wouldn't be worth having.

Returning to our main subject, time, as I have said, had many different meanings for different thinkers before Einstein. I identify four such meanings all of which, due to the importance of time, have become the focus of mass consensus, following, or even civilizations.

First, we have the clever and profound, world-shaking, intellectually brilliant interpretation of time proclaimed by the Irish Prelate, James Ussher that God created the universe at exactly noon on AD 4004, implying that time was already running from the so-called Time Zero, marching on to give us the story of history. For generally history was only understood as the march of time. Let us call this religious view 'Act of Creation'. Even school boys can see that it has not much intelligence to commend it. Yet the majority of mankind, numbering billions seriously believe that this is how the world came to be in existence, and worship anything they believe to have 'Created' it. A religious interpretation of time that nevertheless links it close to life as all the logical analysis suggests. There can never be a final, definitive theory of time satisfactory to all mankind. Properly, we should all live with what we believe and forget about the rest---except that time is a powerful agent of causes in science.

Next, Isaac Newton. Newton believed in absolute space and absolute time, and since he was very great in science, everybody able to think followed the great man's definition of time, and it ruled scientific thought until Einstein. Even still now many scientists speak about time as if they think it's running all through the universe like a thread—or that it's general. To be honest, I really consider those who claim against the facts that time does not exist more intellectually respectable than their opponents, mostly scientists and philosophers, who insist that it just is, whatever they mean by that cryptic assertion. In the words of one book reviewer already mentioned above, "Time does not flow, it just is", a biblical language expressing a biblical myth. For if that is so then what is the use of the earth-year---why should we try so hard to find a logical explanation that ends in the existence of the year? I rather accept Bertrand Russell's theory that it is constructed as "relation between points". Thus sentience is required, because somebody must be there to set the points for the yearly cycle; which means we are sent back to the beginning as the religious people claim that only God or a Supreme Being with infinite powers could guarantee that somebody with the necessary intelligence would be there to do that. So, in the end, we have to realize that nothing in life is easy to explain without an ultimate attribution to a deity, least of all time! Even though we know God does not exist, at least Bishop Robinson told us so in his book, Honest To God.

The third movement is the Lorentz/Einstein denial of absolute time and absolute space, demonstrated (or proved) with scientific experiments. Soon after that we got the fourth interpretation of time as his own mathematical interpretation of the Einstein notion of time, called the Minkowski formula for space-time continuum, or 4-D geometry for short. Its brevity belies its momentous effects on scientists. Because it was supposed to be scientific, or the mathematical interpretation of the Lorentz/Einstein experiments, practically all scientists refer to every time and every space as 'space-time', meaning that space has been equated to time---that the two entities have become one. And yet Minkowski could not define time. Also the very great scientist and mathematician who confirmed the general theory of relativity, as we have noted, Professor Sir Arthur Eddington, the founder of astrophysics, described the theory as arbitrary and fictitious. I have therefore rejected the Minkowski formula as logically untenable, since it is based on imaginary time coordinates. The 4004 Creation of the universe is too dumb to think about, and Newtonian absolute time is abolished by Einstein. My rejection of the Minkowski theory is not mathematical in the sense that it is not written in mathematical symbols but in words. However it is somewhat mathematical in the sense that it is a logical objection and all mathematical statements have to have a logical premise—and it is his premise I am challenging. I am simply pointing out that in the absence of a universal time Minkowski had to show where his imaginary time was coming from; never mind that it's imaginary, but from where? Also how can anything imaginary be relevant in discussions about time?

So we are left with the rational consideration of time Einstein proposed, namely that time is derived from your local space---points together with the ability to count and sentience are required, exactly the way we obtain the year, which is the basic time unit on this planet. For the philosopher, the problem is not that time does not exist since we are using it daily in all sorts of activities, but how we get it, how we 'construct' our time, as Russell put it, because cosmic time is abolished by Einstein. This is the current logical situation about time overall---that is, scientific, philosophical and practical. Now let us look at the specifics, or how some writers have considered the matter.

With regard to the view that time does not exist at all, my understanding of the thesis of Professor Yourgrau's book mentioned above is that the legacy of Gödel and Einstein to the effect that time cannot exist under relativity has been scandalously neglected or forgotten by the world. Additionally he says, when that is given its proper due, time travel becomes a scientific possibility. But this cannot be true because the very idea that there can be such a thing as 'The Logic of Time', as proposed here, is derived from

Einstein's researches; this logic makes time necessarily discrete, so that we count individual years to get the centuries and our own numerical ages. The fractions of the year too (the seconds, hours, minutes and days) are also separate and individual units of time. Obviously, there is no chain in time; every unit is uniquely separate---Professor Whitehead's 'non-interacting moments'. That is what is meant by discrete time. However, discrete time, such as we have here on earth, makes suggestions of time travel laughable. Yet the sad fact is that most of the recent writers on time seem to be only interested in theories that make time travel appear to be, as Professor Yourgrau put it, "a scientific possibility". Or to quote him in full: "Gödel, the union of Einstein and Kafka, had for the first time in human history proved, from the equations of relativity, that time travel was not a philosopher's fantasy but a scientific possibility." Kurt Gödel bears most of the blame. He asserted that his discussions with Einstein had convinced him that time travel is feasible. I concede that scientifically that could be so if time is naturally equated to space; but so long as they are separate entities they cannot 'curve' together as to suggest that time travel could be possible. The Minkowski attempt to merge them with mathematics was not successful---even he himself admitted that before that they're separate entities. And yet, as discussed below, in nature we human beings can never make two phenomena one or one entity two---see App. II.

Thus I wish to assure the reader that there are absolutely no such (logically valid) equations in relativity that could even remotely make time travel seem plausible. In special relativity, there is none at all, and yet that is what concerns us most, since there is nowhere anywhere in general relativity to be capable of giving time there to anybody. The so-called interpreters of general relativity carry earth time there in clear breach of the Einstein theory of frames. What happened is that mathematicians coerced Einstein to incorporate the Minkowski formula for 4-D geometry into the field equations of general relativity, which was very easy. Yet the Minkowski formula is logically flawed and therefore completely unacceptable as the basis for altering physical reality to one of four-dimensional space, in fact, it bears no relations at all to physical reality and looks beautiful only in his own mathematics, which is considered arbitrary.

Let me point out that what the mathematicians wanted was a new formula to replace the 3+1 system (as they do now, even though they know that it is not logically valid), so that they could write one equation to represent time, space and matter as '$s=ct...$' It was an extraordinary demand to address to inanimate nature, and would be funny, something like a silly prank, if it were not so serious. Man is an orphan because he does not know whence he came. He is a clever orphan/ape on a lump of rock; and for such a creature to demand something from nature by way of his own mathematics is nonsense. Yet they went ahead, and Minkowski obliged. However, Professor Yourgrau himself has quoted David Hilbert as saying Einstein did not believe in the concept of 4-D geometry or four-dimensional space.

Therefore space and time remain separate entities, especially on any inertial frame like the earth, as Einstein made them in special relativity.

In all history, time has been regarded as the same as existence, hence the age-old definition of time as 'the irreversible passage of existence'. In effect, it equates motion with time. In addition, since nobody knew (or still have any idea) how life came to be, no one worried about the nature of time which is closely associated with it. Hence the mathematicians chose to refer to time as 'just is', meaning it just happens to be there in nature; and so time and life were lumped together as constituting the eternal mystery on earth. The genius (unique insight) was the suggestion that it does not even exist until you have invented it out of the parameters in your environment. This is the Einsteinian notion that Kurt Gödel misinterpreted. For time does exist on earth; the fact, however, is that it did not exist before we came down from the trees. It means we created it; and since its creation it has come to dominate our minds and everything we do, like water (or language). Nobody is born demanding water in order to live. But once we taste it we realize that life cannot go on without it.

The mathematicians used astronomy (being the natural features of the world, as perceived) to construct clocks, and that was it. We simply used the time provided by the clock makers. This time was taken as general and absolute; it was assumed to have originated from divine sources; and since a centrally imposed time system could not be different in different places, one second here was supposed to be one

What is time?

second everywhere else. Then Einstein burst on the scene with his new idea of time, showing that the old idea was religiously imposed and not really true---obviously time is neither general, fixed nor absolute. He proved this with experiments.

Given the supreme importance of time in human affairs, the irrational view of time (before Einstein), influenced all life and all ideas including religion and historical narratives adversely, spawning mythologies many of which are still with us today, with their own consequences, some of which are even detrimental to human existence. One of these is that history is the march of time---marching since the 'Dawn of Time'---instead of recognizing that history has been the story of how life has been lived through successive events since the 'Dawn of Existence', or intelligence, or the first acts of sentient beings on earth, from which acts all successive events have flowed as inevitable consequences, thus giving us the continuing story of human life on this planet. There was no 'dawn of time'; there's only the 'dawn of existence'. Today, time is not seen as marching on and taking us with it; rather we think we are marching on event by event and recording them as they occur at certain dates and times. These dates and times are the positions of the earth round the sun. A time unit is equivalent to a physical distance round the sun.

(b) - WHAT IS SECULAR TIME?

First of all, (before we discuss time after Einstein proper), we have to define what is meant by secular time. The Einstein theory of time is called 'secular' in the sense that it is traced or deduced consistently from premises based on material reality without any attribution to any god or deities. Thus what has to be explained is the phrase 'secular time', and that is this: everything we refer to as time must be a recognized unit of time. The word time has no meaning without quantification---see Appendix 1, 'Time and Quantified Time' below. People usually mention the word time to mean the passage of anything---events, moments, even sitting still---but that is not logical thought or attempts to define time as it is used in all activities in society. Time in science, logic and life (or what of it that we need to understand or explain in ordinary linguistic usage), is presumed to be either materially based or mathematically calculated from features of the earth logically, as opposed to time that is simply imagined or mentioned as the term for any passing moments. One common example is time in dreams; another is just referring to time in ordinary conversations that are generally understood to mean any moment of passing consciousness---the shortest time, the longest time, and so forth.

In society or real life (as opposed to these vague instances of mentioning time without definition), it is necessary that everything we refer to as time---every unit of time in use---is logically traceable and derived from the

What is time?

periods of the revolutions of the earth and its long orbits of the sun---usually shortened to the phrase 'motions of the earth'. Not long ago the individual units of time, the hours and minutes and seconds, were regarded as divine; God had actually created them as independent entities. They're easily explained with mathematics as fractions of the year that would not exist without it, yet the mischief-makers claimed they had independent existence. Even still now many scientists continue to speak of time units as if they're mysterious. But of course they're not. The year is determinate. All the units of time are also formulated as fractions of the year and its astronomical features so that a certain number of each unit will add up to exactly one year to coincide with the complete orbit of the earth round the sun. Thus there are no thirteen months; no 54 weeks; no 368 days in one year. The units do not go on after the year; they are all recounted from the base of one at the end of the year.

The 24-hour periods and the long orbit of the sun provide considerable periods of planning time for all activities: time to catch a Bus, plane or train; time to eat; time to walk a distance; time for work; time for sports---time for doing anything at all. All of these are derived from either the 24-hour periods or the earth-year. In secular time we realized that a mathematical or logical explanation was required, and our own Bertrand Russell, as the world's most recent great philosopher (who was also a logician, writer of genius and a great mathematician), provided the world with an appropriate theory called 'relation between points'; that's the only time that can be programmed into the clock, and we all know that time in the clock is the only reliable time.

Time, he said, is a construction. Together with his collaborator, Professor A.N. Whitehead, he also interpreted the world of sense as 'a construction' rather than an inference, to overcome the old practice of philosophers inferring all things and connections in their minds as their logical definitions of physical reality, contrary to the physical reality discovered in physics, or the actual physical analysis of what we perceive and can also infer from what is perceived. So secular time refers to the time system we can consistently trace from the mathematical, logical and visual premises all combined. Every mention of the Einstein theory of time is to be understood as 'secular time'---traceable from material reality without mythologies. The only system of time we can program into a clock. It is conceivable that Russell gained his insight by asking the question, if cosmic time is abandoned then what is put in the clock as time? Given sufficient logical acumen, everybody can deduce that the new theory is calculated (and can only be calculated) from the motions and physical features of the planet we live on. That's the meaning of secular time. Yet time belongs to human beings and not the cosmos; it is not part of cosmic reality, as I have already explained that what happens on planets (as implied in the concept and postulates of special relativity) cannot be relevant in the universe at large---and they include time as a conceptual phenomenon used for the control of events that are not recognized in the cosmos for several reasons. A natural law on earth is not in the same class as natural laws in the cosmos---they involve only massive objects, moons, planets, stars etc. Activities on the surfaces of planets, on the other hand, are so many billions and too small in size to matter in the cosmos.

(c) - TIME AFTER EINSTEIN

Albert Einstein changed the debate about time for good with his division of the universe into two distinct categories, governed by different natural laws: (1) General existence, or general relativity, where objects or matter just existed and whirled around under the influence of gravity without any conscious directions or time and order; and (2) special existence 'in' special relativity frames or bodies, where the two postulates and time applied, perspectives arise and intelligence and life can flourish in response to the intelligent use of available resources for civilizations to rise and fall---or generally for life to flourish as it cannot do in the whirling flux of general relativity; thus creating the never-ending chain of events known as history or the continuing story of human existence. Since civilizations arise upon definitions of time as mentioned above, the reader can see that only the very rational, scientific civilization can be consistent with the new concept of time.

Einstein did not deliberately set out to change our view of time. It was an accident discovered by Lorentz. He said the Lorentz concept of local time may be regarded as 'time, pure and simple'. His genius made it sound simple, but it was the beginning of the most profound revolution in human thought. It was unique; the nearest idea pointing to the origin and purpose of life because time is the second most important thing in the universe, bar the life itself. And the two are inseparable. No thinker has any idea as to which is which or which of them came first. Personally I believe everything in human experience is generated by the brain in us; this implies that all human creations are secondary to the existence of life part of which is the brain.

Of course, on the other hand, Einstein did deliberately (and even contrary to classical physics), set out to change our views of the universe. The result was the theory of frames with which we are now familiar. It divided the universe into two distinct categories. One is general relativity, where there is nowhere anywhere for life to evolve and flourish; the other is the inertial frame, where life is possible and civilizations can rise and fall. Time is required in this second division of the universe; and the local time idea was just the thing to suit inertial frames. I think we should now write time as the third postulate to add to the two original postulates of the special theory of relativity.

The problem thence is to discover how our own time began, not as a version of a universal time, but a time limited to our frame; a major kind of philosophical inquiry since time is inseparable from life. That old idea was a mistake; yet everybody in science is still considering time as if it is something generally in existence and our time is a version of it. Thus, the Minkowski formula for 4-D geometry is defined as incorporating time in the three dimensions of space to create 'space-time', the merging of space with time, the end result of which is to give us what we call time as 'space-time'! In the absence of a universal time, where is the time incorporated into the natural dimensions of space coming from, if the end result is only to create time again? Using time that does not exist naturally or universally to create space-time as time by means of mathematics--what sort of logical reasoning is that? I am silly enough to let it bother me a lot; I really do not believe that it worries anybody else since everything I write is never even read in manuscript let alone published---my long suffering son has had to do it on my behalf, yet he's only an engineer! It may well be that people are literally afraid of time---afraid to offend the Gods. Yet, actually, the nature of our time is easily deduced from elementary logic.

In fact, as Russell put it, "There is no longer a universal time..." Thus, he asked, "What is measured by a clock?" Yet the question is wrong. The clock does not measure time. It rather reproduces units of time specifically programmed into it for reproduction. That is the reason it works in units only---second, second, second, and so forth. The real problem is how the units of time programmed into the clock are derived in the absence of a universal time.

The year, of course, is basic. The seconds and all other units of time are derived from the year with mathematics as fractions thereof. Everything depends on the use of points. We use points to get the year. The fact that it is repeated over and over again to give us all the centuries means our time is determinate---in other words, our time is discrete. The essence of a discrete time is ended when the units are expended, thus we have to repeat the yearly cycle for our time to continue all the way to the centuries by replication. In addition, the system runs all through our units of time: from seconds to the minute, minutes to hours, hours to days and so forth. Our time is not a thread running through nature as we used to think; what we have found through experiments is that it consists of a chain of individual units created with points or mathematics in association with astronomy and the essential features of the globe; as such it consists of separate moments, as Professor A.N. Whitehead has confirmed.

It also means time is not known ahead; what we call time, say the year, is known after it has passed---e.g. the year is not ended until 31st December. That is when we can have a whole year. Then we have to start another year. The same principle applies to all the other units of time derived from the year, including, as I keep reminding the reader, the atomic units of time, because they have always to be related to the second to make sense. Secondly, time cannot be seen as the cause of events; events are physically caused; the times are recorded as the periods during which they occurred. Thirdly, time created with points and which is not part of a universal time, cannot have anything even remotely to do with what the religious leaders dream up about the nature of time. Fourthly, the passage of a time system produced with points unit by unit, as the year shows, requires no arrow or arrows to pass through nature: the units replicate to pass by---precisely as the years replicate to become centuries. All that remains for time to take its rightful place in science as a rational subject is for mankind to wean itself from the 'sweet' religious suggestions about time (what Professor Eddington called 'even-flowing time'), since the true facts are now well known: we don't know what it is, except to guess that it is the product of sentience, physio/chemistry and motion; but we know how it begins; we also know how it passes by---second, second, second; or year after year after year; and we know how it will end, that is, when our planet ceases to support sentient beings who can count the orbits of the sun as 'years'. Religion has nothing to do with it. The arrows of time for its passage through nature is redundant; and is definitely not universally existing in the cosmos because without knowing how to count the orbits of the sun as years, there could be no years just bland existence.

In conclusion, let me point out that, if the distortion from the Minkowski formula is eradicated, the question of time under relativity becomes simple, exactly as Einstein put it, namely 'pure and simple'. Here are the basic facts: (1) There is no longer a universal time so we have to search for the origins of our time because; (2) every 'body' or inertial frame has got to have its own time; (3) under relativity the all embracing time is a construction, like the all-embracing space; (4) both the earth-year's time and the atomic time use regular or repetitive motions to track time-

---that means they can only track passing time since the pulses or motions can be counted 'after' they have occurred and not before. We put all this together and get the notion that time cannot be logically defined, which means that what we call time are units of passing time. They are units because we get them from repetitive motions or cycles---and that is the reason it is passing time, simply because, of course, these regular cycles are passing. One after another (or year after another year), there is nothing more to time.

Thus, in the end, since the years are our only means of noting the passage of time, the explanation of time was rather easier than going through all those complicated mathematical and physical theories of arrows, mysticism and divinity. It is conceded that time is mysterious. It is even assumed to be the last refuge of God after Charles Darwin, since many people believe that time's deep, fearful and intimidating mystery goes beyond human comprehension. Yet it is rather ironic that we have been using the orbits of the sun for time without realizing that it is all of our time---mere physical cycles counted as years, centuries, millennia---because we thought we were measuring our version of time out of general time permeating the cosmos, the provenance of which was assumed to be nothing but divine. Yet once we learn that there is no longer a universal time (thanks to Einstein), and that we do not measure time at all, the orbits of the sun appear in a new light: we count the mere physical orbits as

What is time?

our ultimate units of time (the years), out of which all other units are derived.

Chapter Seven: Entropy, Gravity And Time

Concerning entropy as racing with time to our doom, or time causing the increase in entropy to our total extinction, and also the effects of gravity on the whole of time through a single clock, I would say this: it all depends on how time is seen or defined, but I believe the idea that entropy and time are naturally moving all through history to just one sort of destination which is the death of heat and activity so that energy and life will come to an end is religious and echoes the Day of Judgment mythological sermons. Can this sort of thing happen in places like the sun where there are no religious inventors of such theories?

What is time?

Now that we know time is not universal, any objection to the logically deduced concept of the time we have must be flawed or just religious sentiments not to be taken intellectually seriously. Similarly, I think gravity is given far more scientific kudos than it deserves. As important as it is, if gravity affects any clock then such events should be interpreted like the clock paradox, since it is known that acceleration affects some clocks' performance. For it cannot be stressed too strongly that any clock's performance under any circumstances is not the whole of time per se, but that of the relevant clock alone. Regrettably, scientists continue to ignore the fact that under relativity time is not running all through the cosmos in the form of a stream or thread. So there is no longer a universal time, and since that is so time cannot be affected by other forces in nature, since it consists of "non-interacting moments or contacts". Ironically, ordinarily people (perhaps it includes some scientists and philosophers) don't think about time logically. Everybody is in the habit of taking time to be 'just there' like life being just there---it is strange, yes; but it is there for our use, and so we just use it.

Yet they're wrong, but do not know it, but since time is always there for all of us to use, nobody cares much about the small matter of its nature and origins. The truth is, time has not always been there for our pleasure. Like language, we created it out of necessity. Moreover, the religious notion that it is divine---general, fixed and the same everywhere---is no longer credible because of Einstein's researches and the Lorentz discovery of local time, as Sir Arthur Eddington has pointed out. Thus logical analysis of time began some time ago, and we have found that it is also digital or discrete, the reason we can have it in units, which is not that difficult to understand since our time is based on the yearly cycle; and the year is determinate and so its fractions are bound to be in units or digital. From all this we now realize that time is much more complicated than just being in existence like 'Being': Time cannot be equated to 'Being' because it requires the intellectual use of points and therefore human in origin.

A great deal is due to change especially in physics and astronomy because of this idea; scientists are ignoring it because they simply do not know what to think. We need another Einstein. Of course, time appears to be some kind of a continuous thing or entity, but that is due to the extreme smallness and speed of the particles of light and everything else, known as the atom or quantum of contact---almost precisely like the steady images of the cinematography; they are digital but seen as continuous phenomenon, many particles behaving as if they constitute 'one continuous thing' due to the high speed by which they travel. All images are digital, but rolled out at great speed, steady images appear, that's what makes the pictures of the cinema. To my mind, it even helps to refute the Platonic Theory of Ideas---i.e. ideas do not exist anywhere until they're constructed by particles of light. One problem is this: even though time appears to be some kind of a thread running through nature to justify many scientific

What is time?

suppositions in theory only, but since it is not so in fact, some of these scientific theories derived from that concept of time are wrong. They can ignore me for challenging scientific orthodoxy, but the truth cannot be denied forever. There are no conceivable forces in nature that can interfere with an entity that does not persist but only appears in flashes and gone. The second problem is that although time does not persist, it nevertheless accumulates from its individual units; yet these units occur so closely together that they appear to be laced to one another, giving rise to conditions that resemble general time, which is confusing scientists. That much is conceded, but the third problem is that, however closely together, so long as there is space between them, they are individual units (non-interacting units, intervals, moments or periods). All this hangs on the nature of the ultimate units of time which are known as 'atomic units of time', so small individually that the closeness to succeeding and preceding units are barely noticeable---yet still it means they are individual units of time that cannot be susceptible to interference by other forces.

So, then, time is now known to be digital, as Professor Whitehead has observed ('a sequence of non-interacting moments'), consisting of numerous contacts of varying lengths in on-and-off, long-and-short, perceptions because there is no general time; yet, like Professor Yourgrau in A World Without Time, some writers have opted for the easy way out by arguing that there is no time. But we do have time. The task is to find the method or logic by which we get this time.

However, stubborn and undaunted, I have invented an argument for proving that no one clock can control time as a whole, which I have shamelessly used in many of my books so much that I am growing tired of it myself. But let us suppose that in a special country nightfall occurs at 6.pm on the dot---this is not that fanciful because in some tropical countries the sky comes close to darkness after 6.pm. Now, if there is darkness at 6.pm, then any clock running fast (or erratically) and therefore showing the time as 6.pm ahead, cannot make the sky grow dark simply because it is showing the time as 6.pm. In other words, no one single clock can control time. Time is obtained elsewhere and mechanized in a clock; and any clock's showing of it may not accord with the essence of time as a whole which is strictly based on the motions and environments of the planet, which means that metaphysically time itself can never change unless they (the motions and environments) are altered. And even if that were to happen, it would take millions of years for another time system to be created out of the new parameters---as it did in the past; for time is more complex than anything else in the cosmos. To demonstrate how intricate it is, the logicians recommend looking at it as the human life itself, as a functioning human being's life with the gift of mathematics to sub-divide this functioning life (with repetitive cycles like the year, pared down to seconds), into periods or intervals and call them time in

What is time?

units and apply them for the regulation of the same life as it goes on and call the whole contrivance 'time by the clock', and probably divine and try to determine the age of the universe itself with this time. We then go on to ask how man acquired the mental ability to do this---that is, to create the intervals or time units and mechanize them in the clock; yet all this will only bring mankind just a little closer to the true nature of time but never the real thing. Nobody can ever know the true nature of time, because it is inseparable from life and we don't know what life is. The best we can do by means of scientific thought is to suppose that, in a convoluted way due to the complex nature of time, every sentence in this book from the first page is aimed at preparing the human mind to accept that, "Time is a period of waiting based on external parameters as conceived in the mind of a 'Sentient Being' in the form of intermittent durations from activities (touch and go etc, as the seat of time in the mind), and externally mechanized in the clock by means of repetitive cycles---like the year, for instance, pared down to the seconds and atomic units." Always remembering that the atomic units are based on the second, and the second on the yearly cycle, so all time is based on our orbits of the sun. Every unit of time is a fraction of the year. If the yearly cycle did not exist, time could not exist either, and life as we know it would not be feasible. Many mathematicians believe that if the orbits of the sun did not exist we could have chosen something else to give us regular motions for marking time; but I think they're wrong because time is tied up with the environments of the planet which are hugely influenced by the sun.

So, sadly and rather reluctantly, I have to admit that the religions are right about one thing at least. Life is based on time, not on 'time allowed by God', or time decreed by God; but in the sense that without time people would be born all right but die immediately afterwards because the technology and culture to nurture the perilously fragile human infants would not be there. This, to me, is the ultimate secret of life on this planet: metaphysically all life is based on time, and the seat of this time is the sense of duration acquired as an inborn faculty from the womb. Time in society is another matter. We are talking here of layers of knowledge at various levels all the way to 'Creation', or evolution to you and I! Time in the clock was constructed little by little by three classes of mental workers: mathematicians, mechanics and logicians, of whom the most important were the logicians, for the concept of 'individuation' invented by logicians is the most important notion that underpins all scientific thought; that's how we get the points and instants. This time is based on life, meaning we have to come to life and learn many things to be able to construct it; and the great Englishman, Bertrand Russell, was very clever to note that under relativity time is constructed by man. I have said in another book that time controls physics and that physics is going to have to change many theories as the four-dimensional world (4-D Geometry, or the Minkowski universe), upon which it is now based, is logically flawed. In

What is time?

fact, time in both senses above controls everything we do---intellectually and culturally.

I repeat, the theory of time has changed; it is now not absolute or fixed and does not cover the whole universe in one format; and since time forms the basis of life and all activities, all ideas about physical reality since life began have got to change with it. But obviously the definition of time itself is complex, as an intricate matter cannot be defined without complexity, but perception is crucial. Of all that, perception is the one indispensable factor. It is the basis of the sense of time, for without contact or activity, the sense of time, and the need for it, will not arise; and since there is contact in the womb, the sense of time must be inborn, confirmed by gravity under the feet through intermittent activities like walking. But any blind person (if born blind to begin with), who never has any contact at all, will lose the inborn sense of time, because knowledge of physical parameters activate the sense of something coming and going, long and short, on and off, and therefore of the intervals between them (intervals between events) which we know as time intervals, or time for short.

The irreversible passage of existence is not a valid logical definition of time because existence is multitudinous and do not all pass or follow one system in anything: some go up, down, forward, backwards, age through chemistry or even stand still on one spot for centuries, like boulders and dwellings---even mountains. You can use any repetitive cycles to give an age to a mountain standing still for centuries and then say time has 'passed it by'. Yet that time

Samuel K. K. Blankson

happens to be just the number of our own cycles and their passage is what we call the passage of time! So the passage of time needs no theories to account for it, since the passage is just any number of any cycles chosen by us (as time, so the passage of them is the passage of time) and therefore means absolutely nothing in terms of longevity. We know and can know nothing about time except the numbers of our own chosen cycles to mark time, and as the cycles pass by we say they constitute the passage of time. Yet the passage of time is the number of our own cycles--- the orbit of the sun is just one such cycle, and means absolutely nothing (I am repeating these ideas on purpose). We can never define the temporal length of one year in logic; we can only use the year to determine time units as fractions of the year since it is a repetitive unit of time. Otherwise we're forlorn and lost but do not realize it! As a result the age of the universe too cannot be measured accurately by man. There is no yardstick for estimating the age of the universe, since the length of our year is unknown. But given that our sun is a tiny object in the cosmos and circling it does not take long, our year as a measure of ageing applied to the universe is problematic, to say the least.

However, any objection to what we now have as mankind's most logical theory of time (since we learnt that time is neither general nor absolute and that it does not run all through the universe and the same everywhere), should be regarded as either fantasy based on ancient traditions, a religious idea or mythology disguised as

What is time?

rational thought. Einstein is said to have failed to establish the true nature of time because he could not explain past, present and future. Yet he did explain it absolutely clearly. He described it rightly as "stubbornly persistent illusion"; and, as always, he's absolutely right because it is still stubbornly persisting and remains an illusion. So far I think the reader of this book will have realized that time, as defined here, cannot be racing with entropy to our doom, neither can gravity affect discrete time. The truth is that all the ideas circulating about time refer to a universal time, fixed, general and absolute, covering the whole universe in one format. Discrete time is as yet unknown outside the highest circles in science and philosophy, and since time is regarded as mysterious the general public does not get near. They even claim it brings a curse when probed too deeply! The problem is that absolute time is abolished under relativity, while discrete or secular time has been proved in Einstein's researches, as Professor Eddington has confirmed. The jury shouldn't have much problem deciding this issue in favor of secular time.

I must add that in our religious and ignorant pre-scientific existence, when the nature of time was a mystery with all thinkers displaying their 'massive' intelligence by producing arguments to deepen the mystery about absolute time rigidly running all through the cosmos and giving us the story of history and so forth (note that the story of history was more the march of time than the events engaging people), it was perhaps intellectually permissible to regard past, present and future as significant topics in the discussion about time. But it's sheer monstrous and offensive impudence to confront a great man like Einstein, who had only recently discovered after experiments that absolute time did not exist, to explain past, present and future. He's right to reply that it's an illusion. Again I have to stress that what happens to us here on earth do not count in the metaphysical explanations of physical reality because we are so insignificant; in particular our concepts, thoughts and emotions, hopes and fears that we have formed from our experiences on the planet cannot be known outside our tiny heads---the universe is so vast that it can only deal with massive bodies, and even then mainly through accidents and random collisions. Conscious, ordered, time-controlled events and all our sensible actions do not exist in the universe. The universe has no time, because time requires sentience. In general there is no time at all; any time is your own local time---your own 'constructed' local time. For somebody must be there to set the points for calculating the repetitive cycles as time intervals or we would have nothing to reckon time sequences. So it means the religions have been right all along (especially Pythagoras.) The universe does not seem to have time and is best considered in religious terms for the consolation of humankind. Logic does not seem to exist outside the human mind. Yet the religions promote wars---that is the dilemma for mankind, and it is permanent. Religion is human, and all human institutions are defective in parts. The only solution is rationality, but the religions cannot promote rationality undermine their authority. Thus, as I have said, the dilemma is permanent, perennial,

What is time?

insoluble and a cruel curse to spoil life for us. Probably just an innocent twist in inanimate matter, a few molecular quicksteps, but a permanent curse for life on earth. Blaming the religions is no use---they also know not what to do. The universe is clearly overwhelming. Man should always remember that we're all fallible.

All the same, we do have serious problems with time; problems that are leading even good scientists astray so much that they're questioning the Einstein notion of time and preaching that time travel is 'a scientific possibility' while claiming at the same time that Gödel has proved (beyond doubt) that under relativity time cannot exist---a contradiction since we have time and also continue to live in a special relativity metric. Really, some scientists are behaving like a bunch of philosophical amateurs. How can anybody travel by something that cannot exist?

Of the numerous concerns about time I have selected only a few topics for discussion. Nobody except a super human being can answer all the queries, legends and myths of time. But Time Dilation, The Clock Paradox, The Twins Paradox, entropy regarded in science as showing the direction of time and the effects of gravity on time appear to me to worry serious scientists so much that I have decided to mention them here and comment on them briefly.

The order of time has already been discussed. Frankly, most of the issues that bother scientists are not really matters of serious concern. What we thank Einstein for is the permanent liberation from the oppressive restriction on the human mind by absolute time running all through the cosmos from a beginning to a predetermined end. It is this mythical fable that spawned the most intricate philosophies of time from Plato to Kant and modern science, simply because there was no way round its iron grip (enslavement) of the human mind. Even when Bergson said, correctly, that space and time were separate entities, it didn't really register much in the way of the rationalization of space and time before Einstein. If it did, Minkowski would have had to struggle to make any impact, rather than the euphoria with which his theory was greeted to the extent of coercing Einstein to mention it in general relativity---even though we are now told that he did not accept or understand it. One of the best news I have ever heard. Yet, according to Dr. Gribbin, it is still regarded as the best way to understand relativity---if so then since the Minkowski theory is logically flawed, relativity is not properly understood! I know the mathematicians would want to lynch me by the nearest lamp post for saying this, but there you are. That's a deliberate choice. Some people die committing crimes; I wouldn't mind dying in defense of Albert Einstein, for death awaits us all anyway.

(a) I believe Time Dilation is what inspired the Einstein theory of frames of which time forms a part. Even Lorentz saw that. He later confessed that he did not discover the special theory of relativity because he failed to attach due importance to his own discovery: "The chief cause of my failure [in discovering special relativity] was my clinging to

What is time?

the idea that only the variable t can be considered as the true time and that my local time t1 must be regarded as no more than an auxiliary mathematical quantity". Probably it did more. The theory of frames separates special relativity from general relativity, so it is possible they're conceived at the same time. They're certainly connected, and the base was "The dilation of time as a measure of moving clocks", since Einstein explained time dilation to mean that each of the clocks were operating in different frames.

But what is time dilation? It is called "the dilation of time as a measure of moving clocks". Actually time is never diluted or ever dilated by anything except the earth's movements. From time to time the clocks are adjusted to accord with changes in the earth's movements, but the changes are usually so small that only the scientists involve worry about them. So what happened in the Lorentz experiment was pure chance, and it wasn't the dilation of time at all. Through ignorance that is what it was called. This ignorance and confusion was common before the Einstein theory of time. As Professor Eddington has observed, everybody was wrong about the nature of time. Einstein is not called philosopher/scientist for nothing.

There were two clocks involved. The stationary one worked normally; but when the people outside looked at the moving one they found that it was slow, or seemed to be moving slowly. If anything at all this is 'clock dilation' (one clock seemed to be functioning erratically not the dilation of the whole of the time system on earth. Yet this episode has been paraded about that it shows how time dilates with speed, thus deepening the mysteries about time, and God knows it is already very scary to some people. Mathematics can be used to smooth out the differences caused by the moving clock. However, the best solution is the Einstein theory of frames---the moving clock is in a different frame of nature or the universe.

There is no theory or action (even mathematics) of any kind that can dilate or dilute all time from the mechanical functions of one clock alone, for time is physical as well as psychological. The two must agree in visual displays to amount to recognizable time.

It should be clearly understood that our time is based on the earth cycle of the sun; and also all the units of time including the atomic pulses that are based on the second are fractions of the year; you cannot change any of these units without altering the motions of the earth. Again it all depends on how time is defined. And even if Atlas tried with all his might and failed to change the directions of the earth, I doubt that mathematicians can do so with pencils. Their vanity is greater than the might of a superman, I know; but even they themselves have long realized (if they'd be honest) that philosophers helped to advance science more than mathematicians whose basic instincts, as Newton said, are to destroy other people's theories. At best what happened in the Lorentz experiment is 'clock dilation' not the whole of time as such. But as clock dilation, we know that a cock, any clock, does not create time; it cannot produce time units out of the sky. A clock is manufactured to reproduce units of time obtained from a breakup of the

What is time?

year and deliberately programmed into the clock with a specific mechanism to reproduce it in specific units, so that a certain number of these units will amount to one year exactly---and start again for another year. This is the reason we countdown to a new year in dramatic seconds. A second is part of a year. For example, one second to midnight on 31st December is this year still; one second after midnight is next year, or part of the coming year, and so we'd have one second less to go for another new year celebrations.

In conclusion, I wish to make some of the complex strands in the debate about time dilation a little bit clearer---as far as I can. Scientists are convinced that it proves that the passage of time varies according to the speed of the observer. This cannot be true because nobody knows the nature of time to make such a categorical assertion about it. Always when a discovery is made in science the mathematical interpreters set to work. In the case of time dilation they have concluded that time slows with speed---spawning the twins paradox, clocks paradox and time travel, all the way back to the revival of the Pythagorean Transmigration of souls, dogmatic religious sermons, Day of Judgment, et al. Mankind is basically stupid. We never get over the fear of death and allow it to rule our minds incessantly.

Now, initially the interpretation of the Lorentz discovery of t_1 (or local time concept) was that: (a) the passage of time varies according to the speed of the observer; and (b) the total effect is that time slows with speed. These were regarded as discoveries rather than some writer's fallible interpretations. Both are wrong. The passage of time as a whole does not vary or slow according to speed one way or the other. This was an observation of the functions of only two clocks. In science and logic you have to allow for the possibility that something was wrong with one or other of the clocks---not the whole of time per se. That is set with the motions of the earth and can only vary if the earth's motions are changed. That is the reason we have to adjust our clocks occasionally when the motions of the earth changed. For that is the only metaphysical change in the nature of time we know. With regard to time dilation, the carriers of the moving clock saw no variations, but observers from outside the moving vehicle did. So Einstein conceived his theory of frames to account for the variation.

The errant clock was in a different frame because the universe is basically fragmented. It's one of the lucky moments in human history when an insignificant fact can lead to major discoveries and momentous consequences. Lorentz realized his mistake (and said so later on), that either of the clocks in his experiment could falter; but his local time concept was significant. Einstein said it should be taken as 'time, pure and simple'. It showed that time varies according to space or your locality---most shocking and most revolutionary because time was supposed on the highest religious and Newtonian authority to be fixed, absolute and general, covering the whole universe and the same everywhere. It's so mysterious that this religious interpretation satisfied everybody, even though we derived

What is time?

time from parameters set by ourselves, and also it did not exist before we came down from the trees and grew wiser. Thereafter, the new interpretation of time dilation is that it is not variable according to speed but according to locality or local condition, in short according to space in accordance with the theory of frames; for according to the theory of frames, every space has its own unique parameters for "constructing" its own time system---applicable only to that frame. Therefore "There are as many times as there are inertial frames".

(b) The clock paradox is almost similar in concoction and interpretations. Einstein pointed out that only one of the clocks had experienced acceleration---so it was again a matter of an individual clock working strangely, not the whole of time going off in a crazy manner. Furthermore, and this is crucial, even if it's one clock going off in a crazy manner, real time would not be affected because clocks do not control time. They merely reproduce units of time obtained elsewhere (from the breakdown of the earth-year and programmed into the clock.) Of course we only note time by the clock; that is so but only when any clock is working accurately---and it takes time and effort to find that out. I would call for a little kindness to be extended to the old thinkers locked in the stiff embrace of absolute time in that the new definition of time was not known to them.

(c) The twin's paradox. The mathematicians are very pleased with this one; it holds out the possibility that time travel is really 'a scientific possibility', some would say it's already an achievable thing waiting for an adventurous tycoon (another Richard Branson) to demonstrate it out there for all to see. But, as we all know, pure mathematicians are prone to this kind of fantasy---it's a crazy job, only mad or half mad people do it. Otherwise the twin paradox is not worthy of serious discussion and should be taken as just another of those stubborn, persistent illusions. If the clock paradox and Time dilations are correctly dismissed as aberrations then the twins paradox based on them should not be listed at all as one of the mysteries of time that scientists are seeking answers to. The world of learning consists of two separate strands: one is the speculative and the other is the scientific out of which we get our technology as substantially true of the external world. In the speculative branch of learning people are free to think the unthinkable. It is the duty of science to prove them right or wrong---but the speculative is as necessary as the scientific, for it's the only way man can have knowledge, so long as you learn to tell the difference. Unfortunately not many people can do that successfully.

The mathematicians promote the twins paradox as some kind of their own comradely pastime which delights them enormously; unfortunately it causes distortions in philosophy and theoretical physics. They argue that everybody has time in him because there is space in everybody and space and time constitute one entity as proposed by Minkowski. This however is not true. The Minkowski theory is logically untenably. Secondly, although it seems to some people that the body reacts to time, but in reality there can physically be no mechanism in the body

What is time?

ticking time away, simple because any such mechanism would not know what time to copy since there are no standard time units. Many people still believe that there are such units of time as part of the belief in the existence of God, but they're wrong in logic and I hope this book's arguments will be helpful to them.

Mechanically every clock works alone; it's not wired to work in tandem with all clocks; so an internal clock would not be able to keep up with any specific kind of time; otherwise the question is "whose time?" or "which time?" it would be copying? There is no longer a universal time that is the same everywhere. When biologists talk of internal time they are referring to the tendency of the organs to react to stimuli in a variety of periods in accordance with the age and gender of the person. For instance, a child heals quicker (meaning in a shorter time) than an adult, and an adult far more so than an old person.

In spite of all that, the mystery-makers called mathematicians claim that we have internal clocks; these are linked to space as Minkowski suggested, so they have effects on the ageing process. They then jump to the other, more famous, fallacy, namely, that time slows down with speed, and conclude that as time dilation and the clocks paradox slow our internal clocks down we are bound to age slowly---hence the twins paradox. But physics has not progressed to the extent of sending men to the moon and back and become so fearful with the bomb and other weapons by propagating mythologies simply because they suit the fancy of mystery-making mathematicians working as closet magicians. Science loses its logical purity when ideas are promoted for their mathematical beauty rather than objective truths. To put it as nicely as possible, human beings were existing long before time was invented for the clock; and we've not yet gone through the necessary surgery to implant clocks in our guts.

(1) The ageing process is not wired to move in tandem with a time system that is known to be based on the motions of the earth so that, as far as the ageing process hidden in mankind is concerned, this time is external and even unknown. The phrase 'internal clock' does not refer to any particular system of time physically planted in our guts, but merely refers to the fact that our internal processes can be set to external time to indicate how long they last, or how often they occur---very useful in medicine not metaphysics. The famous Circadian Rhythm refers to how the body adjusts its activities (rhythms) to the world's alternating day and night system, or how the body's functions are affected by the day and night revolutions of the earth; and we can all agree that they're bound to be affected or influenced by the earth's revolutions. After all the whole idea of having a logical time system is based on the Day and Night system. That is why time is practical and not metaphysical, for the Day&Night system is irrelevant in the cosmos or the interpretation of it. Again, what happens on the surface of bodies (and chiefly in heads of the individuals of their inhabitants), are not known in the cosmos; the cosmos reacts to only massive bodies. And there is no sense out there to figure things out---only

What is time?

touch; and our concepts cannot touch bodies in the cosmos physically to be noticed.

(2) There is no longer a universal time that is the same everywhere, so there are no standard units of time in people wired to accord with external time and the mechanics of space travel. But, obviously, if one lives in the tropics he or she has a different biological system, even (dare I say it?) a different skin colour. What happens to the outside of our bodies are reflected inside us; so the day and night system affects how our bodies function, and since our time is based on this system, some activities follow the syndrome and appear to be time-controlled. A single paragraph in a good book on biology will make this clear. The phrase 'Body clock' refers to how the body regulates its activities to accord with the regular Day&Night system. Matters concerning clocks are not questions disputed in philosophy, so long as they are not labeled as 'time'. They will show our system of time (demonstration of time) but do not constitute the time. The time is obtained elsewhere; the clocks are programmed to produce certain units of time to accord with the orbit of the earth round the sun. The clocks may do this mechanically accurately or not; but their errors and malfunctions should not be attributed to the entire time system---hence the clocks paradox does not affect all time. Biological and body clocks are proper subjects for scientific study: the body clock shows how the body reacts to the Day&Night system; the biological clock shows how the different organs in the body regulate their functions by their own rhythms, such that the liver and

kidneys have different periods for reacting to poisons, and so on. As I have said, little by little many of the mysteries of time are being deciphered in science and philosophy; and even that science and logical thought cannot reach give us no grounds to suppose that time is in any way divine. It does not mean anybody is claiming that time is no longer mysterious; it is, but so are the function of the liver, the kidney, the inner ear, the eyes and so forth. Nobody can explain why they have to be there except that they form part of the body to enable it to function properly---yet to what effect since we'll all eventually die? All we demand is that nobody should go about preaching unscientific sermons about anything in nature and impose dangerous rituals on human beings, because nobody knows anything for sure. As far as time is concerned, like many things in life, we have eventually got a fair idea of time's origins and nature through scientific researches. That is the reason for Eddington's severe condemnation of those who still think time is flowing through nature like a stream.

(3) By the same token, (since there is no standard units of time) it'd be impossible for a clock inside us to accord with clocks outside the guts of man. In any case, the Minkowski theory upon which this theory is based is described as arbitrary and fictitious. Science is impersonal; we have no means of altering its conclusions; but there are adequate methods for neutralizing or mitigating most of its ill effects. They do not include falsifying results to make them (religiously) palatable. That won't move a tram, let

What is time?

alone the rockets to transport man around the solar system.

Chapter Eight: Entropy And Time

The theory is that increase in entropy leads to the dearth of heat and activity and that this will eventually result in our extinction; and yet as time goes on entropy is inexorably tending to increase; so in effect we are heading to our doom 'as time passes'. This implies that time is running all through the universe as an independent entity; but under relativity there is no longer a universal time. Nevertheless, according to the former editor of NATURE, Sir John Maddox, the very pillar of the scientific establishment," the reality of the process is beyond dispute". However the theory overlooks the logical fact that time is defined as 'a period of waiting', and that it is caused. Anything that causes a period of waiting is time. I use the term 'processing' for our experience of time. Several agents are suspected to be causing what we know as 'time'---

What is time?

chemistry, inertia, delayed-reaction (such as is common in earthquakes, tsunami and other natural events), or even tugging from dark or invisible matter. Otherwise time does not exist as a physical entity, and Professor Eddington, too, has strongly castigated all those who still believe that time flows, as quoted in this book. So time as an independent entity passing by is logically untenable. Therefore entropy cannot be increasing 'as time goes on'. Time does not flow or go on. Whatever is causing the periodic increases in entropy is not time. Quite possibly they are periodic chemical reactions and nothing to do with the flow of time on earth simply because time does not flow on this earth. It also does not go on---there is only one day; the blips that cause successive days may be useful to us but they do not count in the interpretation of time because they are not natural events but freakish; that they seem to last long enough in our eyes is due to the fact that we are so tiny and our perspectives are unrecognized in the cosmos. We have not yet got used to the reality of life: namely, that Human Beings either en mass or as individuals are not worth a sputum in the cosmos. The very earth we live on does not count, yet it is huge to us. The sun is like a fly yet it is hundreds of thousands of times the size of our' huge' earth---so who are we? Absolutely worthless in the consideration of things like time that go to the roots of our existence and therefore must have relevance in the cosmos. And there is nothing we can do about this sad situation. The electronic revolution will not help; in fact it may end up driving us back to the religions.

Time does not move; it is discrete and the procession of its units (like the year) add up to create the illusion or the false sense of time's passage and continuity. It is even a surprise that the very scientists who swear by the Minkowski 4-D continuum insist that time is passing by as an independent entity, and that as it does so it causes increase in entropy (see the chapter 'Inventing Time' in Sir John Maddox's book, What Remains to Be Discovered.) He asserts that, "An ordered state spontaneously becomes disordered or the entropy is spontaneously increased...Does not behavior of that kind provide an objective sense of the direction in which time increases that is independent of sensation in our heads?" The simple answer is no. Time does not run through nature; it also does not pass, only the cycles we use for tracking it are passing or moving. Time under relativity has no direction. There is no such thing as the direction of time; there is only the replication of time units---the one year is one unit of time and it becomes years as the earth continues to orbit the sun. The year does not need direction to continue. It is only one, and replicates to become many, centuries. Time is discrete because it all depends on the discrete yearly cycle; all the other units of time are fractions of the year, or are derived as fractions or sub-divisions of the year with points in association with astronomical features of the earth. The earth-year is the beginning and end of earth time.

It is common knowledge that scientists disparage philosophy. No one can blame them when even Sir Karl Popper has confessed that he's not proud to be called a philosopher due to the silly and irrational ideas now paraded as philosophical.

Yet still philosophy is unavoidable, and the whole of science needs it. In his review of the book Science, Perception and Reality, Professor Herbert Dingle said exactly the same thing in defense of philosophy.

Scientists are narrow specialists; they need the philosophers who survey the whole scientific field, from biophysics to astrophysics, the tiniest to the gigantic, the invisible to the massive stars so huge that millions of our tiny sun will find room in them; and they need these generalists to ask pertinent questions that the specialists may have missed, for we are in the dark about the universe and why we are here with the ability to ask such probing questions such as this: "If cosmic time is abandoned, then what is measured by the clock?" What gives the human brain its incisive powers is the biggest conundrum of all. At the end of his article mentioned above, Professor Dingle had this to say: "...Science itself is rapidly transforming at least the external patterns of our lives, but those who practice it cannot reflect adequately on what they are doing simply because they have no time. That is an independent inquiry, and a most urgent one..."

Yet, in my opinion, the present set up precludes that. Specialists advance theories consistent with formulas passed on by other specialists; if there are errors in them they get transported to other specialists to pass on till disaster or a major breakdown in communications brings a whole discipline crashing down. Often nobody gets hurt (the general public may not even hear of the problems in certain fields), because in theoretical matters, the distance between theory and practical effects can be so remote as to appear to be none-existent. Moreover there will always be proponents and opponents to any suggestion (the protagonists), firing their heavy guns at each other. But scientific research covering so many fields require the philosophers to knit them into a coherent story to satisfy the human soul and the specialist scientists cannot do it.

As a result it is extremely difficult to challenge any theory once it's been published, no matter how wrong it might eventually turn out to be. Time dilation is a notable example. It's supposed to mean time runs slowly with speed, but obviously that is not the case---at any speed who can read the speeding clock? Those travelling with it are also speeding with it and notice no problem. Before the dust settled on this conundrum, up popped the Minkowski theory of four-dimensional continuum, an idea so seductive that mathematicians just fell in love with it and overlooked its arbitrary nature. Succeeding generations of scientists have used the Minkowski formula to vitiate their suppositions and speculations about black holes. The fact that Einstein mentioned it in his general relativity is no excuse; we now know that he did not even bother to understand it. The answer, I suggest, may be found in the critical examination of formulas before incorporating them in new ideas, although I know that this is easier said than done.

What is time?

But it's sad that whenever they try their hands at philosophy (some) scientists, though at the top of their fields of study, nevertheless manage to utter such banal ideas as to make one feel sorry for them. For instance entropy cannot be marching with time leading to the death of heat, energy and the end of all activity as predicted in the second law of thermodynamics. There is no such thing as 'the march of time'. Our time is discrete---the year for instance---and so it cannot march. I don't know about the death of heat and all activity, but does that mean the end of the universe of all action and motion through entropy? What about the theory of frames? Are we to understand that our tiny planet is so pivotal in the universe that if the death of heat occurs here it will reverberate throughout the cosmos? Anything causing anything in this vast and unfathomable universe is just one instance of the multitudes of events and activities in the universe (positive and negative, forwards and backwards), and it is so vast and complex that only a fool would dare to make absolute statements about it one way or the other. As either individual persons or even en mass and together with our planet we're not even as big as one thousandth of a small ant's leg in comparison to the size of a cosmos consisting of about one billion trillions of huge stars as astronomers are now telling us. Many ideas about time are religious. The end of everything idea, when it occurs in any theory, has Heaven and the fear of death hidden in it or written all over it.

To those who are using these dubious ideas to claim that Einstein's theory of time is untrue, what I have done in this book is to point out that there is natural time but it is caused, materially caused, and is not leading to the death of anything because it is not everlasting. Certain conditions spawn (or enforce) periods of waiting of various lengths on our senses; they also give periods of revolving lengths also implying time; and so for creatures with cosmically brief lives as we are, we can accommodate our lives within some of these periods of waiting—which we know as 'time', or time-span. Thus we use the orbits of the sun or atomic oscillations. Atomic time is also, like the year, our means of showing how time is passing and never what it is (I always have to emphasize this.). But all this is destined to disappear altogether when the sun decays. They have no more influence in the cosmos than a solitary drop of water has on the size and depth of an ocean.

Mistakes in science are almost entirely caused by formulaic thinking---or standing on the shoulders of giants. We've got to make sure they are permanent giants, useful giants, reliable and intelligent giants, not mere religious dreamers in exalted positions in the academe and elsewhere.

Chapter Nine: Gravity and Time

Scientists swear by the Bible that gravity causes time to slow down. Numerous theories have been built on this notion. But, again, it all depends on how the time is understood or defined. One basic thing in the world is that our time is based on the regular orbits of the sun by the earth. This makes our time discrete---consisting of one year at a time, and the year too is pared down to the seconds and other units of time as fractions of the year, and therefore also discrete. That is the reason the units of time we have are all in individual units not in a chain---second, second, second and so forth. But, obviously, discrete time cannot be affected by anything---wind, rain, and sunshine---except the earth itself.

Technically discrete time does not exist as a living, pulsating entity. It's one moment and gone, one year and gone, and the year pared down to moments give us all the other units of time. What is confusing people is that the months and weeks and years are not temporary 'moments'; they last for various lengths of time. This is what we mean when we say planning time is spread over longer and longer periods. But the basic cause of time is momentary---the year is just one moment although it gives twelve months for planning life's activities. It is momentary in the sense that it is determinate and has to be repeated to continue. "A time system is a sequence of non-interacting moments". It is a logical moment, although there may be several months in one of its moments---several minutes in an hour, several hours in a day and so forth. So long as the year is determinate and has to be repeated to continue, the time system derived from the year cannot be anything other than discrete time, with all the momentous implication of such a time system—it cannot march, curve, or cause the story of history, and so forth. It is also automatically secular and logically structured. Altogether, we are trying to conceive a time system to replace cosmic time and discrete time seems to fit the bill. But all theorists are fallible and cannot claim omniscience, except that what is logically unassailable cannot be rejected with illogical ideas. This is a perennial problem. We saw it in the discovery of special relativity. Many theorists could have discovered it or come close to it, but they're hampered by so many formulas relating to the eather, and nobody could tell anybody that the eather debate was logically flawed. Everybody agreed that electromagnetic radiation had to have a medium of transmission. Einstein had to ignore the eather completely to discover special relativity.

The theory regarding gravity and time, as stated by Professor Bernstein in his book, Albert Einstein and The Frontiers of Physics (Opp. Cit. page 110) is this: "In the absence of gravity, space and time are distinct entities. In the metric of special relativity they play distinctive roles..." This is what we want to hear because we live in a special relativity metric. The contentious issue is what he went on to state: "But in the presence of gravity the metric is altered, and space and time become mixed up with one another. The metric has four coordinates, but the space and time coordinates become entangled..." Yet, space and time are not different streams of many rivers to become mixed up, unless time is regarded as running through nature like a stream. The instinctive tendency to regard time as an entity 'flowing through nature' vitiates all the suppositions of all writers dealing with time, except the putative 'three' of Professor Eddington who understood relativity.

The whole matter is hopelessly confused and confusing. If time is not synonymous with motion this cannot happen, and we know that our time is not synonymous with motion because it is discrete, proceeding unit by unit. It's not the same as 'Being' either. That would be a universal time but there is no longer a universal time. We construct our time with points as applied to space to get the time sequences or intervals as 'relation between points'---one of Bertrand Russell's discoveries. Logically there is no other way, the reason I call it 'the logic of time in the universe', applicable to any Beings anywhere in the cosmos. We have had to add something (like the intellectual use of points) to being, or just being there, to get our time, especially in units. The time was not 'given' to us. It is not divine, even if divinity is not nonsense. We could not have got this time without sentience or intelligence. This idea of using points to get time is not that easy to comprehend. 'Being' appears to be time in essence or so closely associated with time that, as explained in the Preface, there can be no human activity that is not set in time; so the very existence of life, our being, or the being of things, exist in time. Yet, logically, being is not time. We have had to add the use of points to get time that we can use and program it in the clock---that, of course, is what we know as 'time'. Being is chemistry. However, taking time as contact with nature, the sense of time must begin from the womb. So we are born, as 'Beings', with the sense of time in the mind so closely associated with our existence that it appears to be part of our being, except that to use it by means of the clock, we have had to learn over several centuries to link this sense of time to external parameters in order to create the necessary mechanism for putting the time sequences in a clock as 'constructive' or 'structured' time.

For we know that time in the clock evolved from very crude attempts to the sophisticated Rolex watch. The result of this evolution is that we have come to realize that our

What is time?

time, being based on the repetitive yearly cycle, is essentially discrete---year after year after year, or second, second, second. Every second is a fraction of the year, so also is the atomic pulses based on the second. This time is not running all through the cosmos in the form of a thread that could be twisted by any force, gravity, or whatever. The time consists of units of time in procession otherwise the passage of time does not occur. The phrase 'units of time' translates into 'units of contact with nature'---that's the meaning of Professor Whitehead's definition of time as 'a sequence of non-interacting moments'. Time is man's relationship/s or contacts with nature. How long they last (which is the purpose of any theory of time), requires a mechanism. Our best mechanism is the clock. However, like everything else, clocks can't fail to function differently under different conditions: gravity, acceleration, etc. This area of philosophical reasoning is difficult, being the apex of human reason, demanding intellects as possessed by Einstein, Russell, Eddington and Whitehead, and shallow thinkers should not demean themselves with childish remarks about matters they cannot understand. There are massive amounts of ideas to ponder in religion, logic, metaphysics and science, as to whether or not time exists, is running all through the cosmos or is discrete and proceeds unit by unit----and how such units are created if they are not divine in origin, following Einstein's demolition of absolute and general time. Everybody agrees that these are the most fundamental questions about human existence overall.

Hence if the Minkowski attempt to equate space to time is rejected as arbitrary, then there is no way that time can be influenced by gravity, however strong, so long as the time is not conceived as running through the universe. So Professor Bernstein is mistaken but he is only repeating the scientific consensus, namely that time can be mixed up with space. Rather the truth is that the time can only come to exist as the product of space and points applied by a sentient being; anything else that resembles time sequences is (or are) caused by chemistry, natural/random causality---something like natural law---or just the way things tend to behave in nature some of which may be beneficial and others are not. And we know this, of course we do. We know this as the natural course of events in the universe when we do not do religion as an act of wishful thinking.

Nevertheless, we are told that experiments have shown that clocks sent to space run faster than those left near the centre of the earth where the gravity is strong. If this is true then we have to interpret it as Einstein said in the case of clock paradox: one clock is working erratically because it is affected by acceleration.

In the end, we have to consider seriously how to do scientific research standing on the shoulders of previous thinkers. Innocently, the famous editor of NATURE was led astray by formulas. He thought (honestly) that the process of time causing increases in entropy is beyond dispute. That is what the academic formulas say. But a philosopher would dispute that assertion for if time is not flowing through the universe then this cannot happen. In any case we cannot be blamed. We did not cause it. We did not cause our time to race in the cosmos and so catch up with entropy.

If under relativity time is limited to a frame and based on (constructed with) repetitive cycles so that it is necessarily discrete and not running through nature like a thread, then it cannot race with entropy to increase it or otherwise. It has nothing to do with it. The academic formula that it does ignore the fact that time is no longer a universal entity racing through the universe to trigger entropy. We may never know the cause of entropy. The universe is too big and complex for us to know much about how it has managed to persist through deaths and renewals, despite these entropy increases. Again, time does not seem to move at all. Properly defined after relativity, time is contact with nature, and what makes it seem to be moving are the repetitive cycles we use to determine how long the contact lasts---i.e. to reckon time. If I have to repeat this a thousand times I will, for it is meant to replace a basic instinct, and that is no easy task. There are numerous concepts we are trying to explain in metaphysics, yet they're not part of the fabric of nature. There is only one day and only one year in astronomy. Days, years, months, hours and the centuries are mere human concepts we have invented (mostly through mathematics), to help regulate live on earth. They do not exist in nature, and are almost impossible to eradicate or correct because we fear live cannot continue without them.

In sum, a clock does not manufacture time as if it is creating it originally and has the power to enforce it on the whole world. A clock, any clock, merely reproduces units of time it is programmed to produce, and whether it does it accurately or not has absolutely no effects on anybody except its owner or owners alone. If all clocks could affect time, then given the millions of mal-functions of clocks all over the world, we could be having accidents upon accidents due to the mal-functions of technologies based on time. Above all, now that we know time is discrete and not running like a stream throughout the cosmos, I personally cannot understand why scientists ignore Einstein's solution of discrepancies in time as caused by the acceleration effects on some clocks. The point is this: if time is not cosmic and we have to trace how we get our time, and we have discovered that it is based on the orbits of the sun and therefore discrete, even including the pulses of atomic time, how else can we explain time dilation? It's like mixing logic and mysticism together and calling it the science of time---we just can't do that. Secular time is discrete because it is based on repetitive cycles; and discrete time cannot run through like a stream to be affected by anything. Otherwise how can any clock's dilation affect the whole of time world-wide? The correct statement should be that it will affect the time calculations of the one using that dilated clock, for it is a case of 'dilated clock', not 'dilated time'. Nothing can dilate time unless the motions of the earth are altered, for every unit of time is a fraction of the earth year, the atomic pulses included because they are based on the second. Even without using

What is time?

the earth year, we would have had to use another repetitive cycle to reckon time----there is no other way. We count cycles as the units of passing time. Otherwise, as our contact with nature, time does not move, neither could the workings of one clock control all time per se. Time dilation is, in reality, clock dilation.

Chapter Ten: Philosopher/Scientists

Plato allowed for a creator. His theory of Ideas is one justification for a creator. But he was wrong. A Theory does not create reality; it can only reflect it through physical evidence. In the absence of such evidence, all theories are matters of opinion. Every suggestion from a human being without physical proof should never be accepted as worthy of attention. Plato has had adulation from religious people for far too long. Even Einstein never got a fraction of that kind of intellectual adulation; yet he rather had the necessary physical proofs. Human beings have been governed by some people's mad dreams for too long---chief among whom is Plato. In Hellenic times, as opposed to

What is time?

Plato, Lucretius rather was right. The quantum theory has proved him right: particles of matter are continually on the move even within a single atom. As they do so (or 'swerve') they accidentally cause chemistry; one result is life. Those academic philosophers still writing footnotes to Plato should be ashamed of themselves as Sir Karl Popper has observed. In what follows, I have quoted passages from scientific writers and others from philosophers and have contrasted them to show how some philosophers (who are not followers of Bertrand Russell) continue to indulge in the fun of bypassing scientific activity as if it does not exist.

In the past everybody thought philosopher/scientists were born not made, but since relativity, the quantum theory and the electronic revolution, the internet, the computer and all the rest of it, every aspects of life has become so complex that we need to train our own philosopher/scientists to guide us through the maze. Socially too things are getting out of hand, so we have to be careful or life can be extinguished very easily through the actions of a few mad men.. The artificial, classroom philosopher/scientists we can create may not be as brilliant as the originals, certainly not, but equally we cannot go on like this without guidance---for instance, relativity is still no properly understood.. Mathematicians claim that it is only with the Minkowski formula they could make sense of it; on the other hand logicians insists that since the Minkowski formula relies on 'i' or imaginary time, it cannot be used to determine the nature of physical reality as to whether or not the physical world or physical reality is four-dimensional.. So what is happening now is a scientific anomaly because all science is based on the concept of four-dimensional space, as space-time, yet logically the notion is fatally flawed. I count myself among the very few people who are writing against it amid mockery from The Royal Society's administrative minions.

Already we are able to train scientists and philosophers; even self-educated people have to learn their ideas from books; that is also training of sorts, except that self-educated people have to work harder. All educationists know that what a professor can easily impart to his students in one lecture would take the self-educated person ages to glean from books. Nevertheless, we train all our experts (physicists, astronomers, mathematicians, biologists, bio-chemists, architects, engineers and dozens of others in technically demanding professions.) They don't come from above; training makes them what they are or what they turn out to be. Many of them go on to teach others in schools, colleges and universities. It is true we don't know how to train people to make discoveries and inventions or propound profound theories about the universe and the world we live in. In a word, we do not know how to create geniuses. Still we do from time to time get some people to whom discovering something out of the usual or from thin air involves sacrifices they would readily subject themselves and their families to without thinking of the money, fame or even the preservation of their lives. From such persons we get our ideas, useful ideas, I must stress, and so I will quote three passages about scientific

What is time?

facts and other three about philosophical ideas imparting great wisdom to illustrate what knowledge means to us. This is a book about time based on relativity, so I think it is right to demonstrate the merits of the scientific outlook as against the Platonic mysticism.

First, a scientific piece (not a mythological one in ancient man's style) about the sun itself as the source of life on earth: "Most of the energy of the sun comes to us in the form of light. Sunlight is transmission of energy, in the form of electromagnetic radiations from the sun to the earth. When Stephenson's first crude steam locomotive was moving along its wooden track, the inventor asked one of his companions what was driving it. 'Your engineer from Newcastle, I would say.' 'Wrong', replied Stephenson, 'the sun is driving it.' I suspect that many millions of us who race the roads in our automobiles have not yet grasped the meaning of this simple sentence. Most of the mechanical work of the modern industrial world is done by energy stored in fossil fuels. Other power comes from water lifted aloft as it flows downwards to the sea...Visible light is the most familiar form of the radiant energy that reaches us from the vast and distant sphere whose surface temperature is estimated to be about 10,000F..." These statements are all facts, even though they may sound strange to the layman. But how have they been pinned together? The answer is bit by bit over several centuries and by numerous contributors. The important thing is that they are true or cannot be refuted without further research. The research will have to be scientific; you could not do it sitting comfortably in your armchair. But equally you do not need to go into a laboratory to do the necessary research. It is a question of attitude. Your thinking must be scientific as stated by Bertrand Russell in his essay "The Rise of Science", namely: "It is not what the man of science believes that distinguishes him, but how and why he believes it..." How he believes it is more important---how he believes what he does, what actions did he take to lead him to those concepts? It is reported in the Press that many people were surprised by the success of the recent science-based stage play Night of 200 Billion Stars, but I wasn't. Rather I was delighted.

After a hundred years of Einstein and several years of encouragement from Bertrand Russell about "The Scientific Outlook", followed by the computer and internet, mobile phones and all, it would rather be surprising if somebody did not write a science-based play of the sort and get acclaimed for it. For science is becoming popular, particularly through space travel, as many mysteries of the universe (or a few of them!), and also wonders of the sub-atomic level are revealed, especially with the magical qualities of the quantum (or a few of them) thrown in. Many books have even been trying to claim that, because of the Minkowski equation of space with time from which they assume the existence of something called "curved space time", time travel is a "scientific possibility"; and although they are wrong because the Minkowski theory is false, the layman is bound to be intrigued.

What is time?

It's not very clever describing science as just one of the many equally valid ways of looking at the world without scrutinizing why, how and what the scientist believes or examines the world for. His job is the logical and systematic search of what is really there in nature and which can be employed to serve the life of man to make it comfortable, longer and happy; for although death awaits us all, scientists want to make human life longer, safer and more comfortable and happy--- by and large, the evils of science come from politicians (some of whom are the most devious of human beings), not the scientific researchers. And one thing that the religions and the anti-science brigades forget is that science is progressive; we wrestle scientific or dependable knowledge form barren, even hostile, nature. Take scientific medicine as a prime example. Obviously it evolved. Man did not know a thing about scientific medicine when life began. It only gradually and even painfully evolved from the researches (and the research methods had to be learnt) of countless individuals, many of whom never lived to enjoy the fruits of their labour. So now that we have grown to know scientific medicine, go to the moon and beyond in search of more scientific knowledge to use for the improvement of life on earth (for instance, astronomy may seem to be a mere academic exercise, but it is from astronomy we are learning how to deflect or explode asteroids likely to end life on earth), we can now eradicate many diseases including polio and small pox; we have also invented the computer, the internet and wireless communication. So, I repeat, it is not very clever to say science is but only one of many equally valid ways of looking at the world---it may be so, but only systematic, logical, scientific knowledge brings lasting benefits. In that sphere there is no rival to science. Besides, nobody knows what is 'the valid' way of looking at the world. Even scientists make no such claim. Their method is to use our most incisive organ (the eye) to observe and report what they see and use them for human salvation. They are not claiming to be capable of doing more than this---but it is enough. It gets us most of the things we need for normal life.

Next, I quote from the popular book, Human Situation, by Professor Macneile Dixon: "In the great arch of night above our heads about five thousand stars may be seen by the naked eye. In their marchings and counter-marchings they make a brave show, yet are in fact scarcely so much as a swarm of bees in all Asia, a spray of blossom in the limitless abyss, where 'a hundred thousand million stars make one galaxy, and a hundred thousand million galaxies the universe'. The stars we see are but a handful, and their removal would not disturb by as much as a decimal the calculations of the angle of their courses. We may be sure that for every human being in the world there is not one star apiece---there are ten thousand. Viewed from the bodily angle, no comparisons can express the insignificance of man among the cosmic magnitudes upon which our astronomers exhaust their eloquence...The earth is a mote of dust, and the sun itself a diminutive firefly. We inhabit the puny satellite of an inferior orb. There are millions of stars so immense that room could be found for millions of

What is time?

our petty sun in one of them." This was written nearly a hundred years ago. We have to update the figures. According to the Philip's Concise World Atlas, p.3, "At least a billion galaxies are scattered through the Universe, though the discoveries made by the Hubble Space Telescope suggest that there may be far more than once thought, and some estimates are as high as 100 billion. The largest galaxies contain trillions of stars, while small ones contain less than a billion." If we can train researchers to establish such complex and yet accurate details of the universe, then we can also train them to do the other parts of philosophers' work, for this is also philosophy.

Scientists or astronomers specifically, can now speak with greater authority about the universe than philosophers. To put it another way, what used to be the exclusive preserve of technical philosophy, that of telling us the nature and composition of the universe, as a matter of speculation or inference, are now displayed openly in elementary books by astronomers with proofs or all the necessary physical evidence required.

For our third example of scientific truths, as opposed to philosophical speculations, on the understanding that man requires both for his material and intellectual advancement, I quote from the New Scientist: "QUANTUM electrodynamics is arguably the most successful scientific theory there has ever been. With stunning precision, it explains the interaction of electromagnetic radiation (including light) with electrons and other charged particles. It is on QED that quantum chromodynamics, the theory of the strong interaction, is modelled." This is the solid ground upon which Quantum mechanics is built. Two Nobel Prize winners put the same idea in different words. In case the Magazine presentation strikes any readers as down-market, I will presently quote them for their satisfaction. First, Professor Richard Feynman began chapter 4 of his book, QED, saying: "...I am going to talk about problems associated with the theory of quantum electrodynamics itself, supposing that all there is in the world is electrons and photons..." and the wonderful Louis de Broglie also said: "Without the Quanta was not anything made that was made." Ordinarily we know light as immaterial. These statements are not only shocking but bother on the ridiculous. Yet they are true, and philosophers who argue against them rather make themselves ridiculous. The religions may object because they deliberately like to object to scientific progress as it undermines their beliefs and make them the laughing stock of modern man---and they will tell you that the more they are mocked the more they like it because it tells them that they are having some effects, whereas they know that religious talk should not be heard at all by people dealing with true knowledge. But a philosopher, as a man of learning, the lover of wisdom, cannot contradict science with mere assertions derived from his arm-chair conclusions. He must incorporate the discoveries of science in his thoughts. There is no way he could do that unless he learns to become a Philosopher/Scientist, which refers to somebody who reasons by means of scientific facts only. So let's go on to show examples of philosophical sayings that fail this test.

What is time?

First, I quote from the Oxford philosopher Professor William Kneal's book On Having Mind (Cambridge, 1962.) In concluding a small book about how and why we have minds, he wrote: "We must retain the Platonic notion of mental events which are distinct from anything in the physical world and manifest a special kind of connectedness." So, according to this 'learned' professor, there is a non-physical world in addition to the physical one we live in, and because of that the Platonic notion of mental events (which deals with that mythical world), is the true theory of how and why we have minds.

In fact, the quantum theory undermines the Platonic idea. The invisible world is that of the quantum, namely images of things can be cut if the lights from any object do not reach the eye. Bishop Berkeley has already proved this without even knowing it, and I will come to that in a moment. But his 'proof' allows us to infer that light radiation from objects conveys the images as the surface silhouettes of objects for us to see them, for we know that the particles of light, the photons, are naturally colored. In an essay on Bishop Berkeley, in his monumental book History of western Philosophy (which I urge the reader to consult about this matter), Bertrand Russell wrote: "Berkeley advances valid arguments in favor of a certain important conclusion, though not quite in favor of the conclusion that he thinks he is proving. He thinks he is proving that all reality is mental; what he is proving is that we perceive qualities, not things, and that qualities are relative to the percipient."

My next example of how philosophers say things about reality to contradict (and therefore reject) what scientists actually find 'out there' is taken from a review of the book, "Science, Perception and Reality", by Professor Wilfred Sellars, an august Harvard professor of philosophy, a man of learning or of wisdom. Only it turns out that what he knows is scientifically rubbish and not worth a farthing:"Professor Sellars nowhere states the purpose of his book, and, since it is a collection of independent essays on a variety of topics, its intention can be judged only from its title. It is therefore not unfair to take it as an attempt at a philosophy of science...From this point of view, its value is slight. It reverts wholeheartedly to the Mill type of bypassing scientific activity, and analyses questions which are quite independent of anything scientists do. To take but one example: 'Philosophers have been fascinated by the fact that one can't have the concept of white without being able to see things as white, indeed, until one has actually seen something white. But this can be explained without assuming that sensation is a consciousness, for example, of white things as white". The reviewer, Professor Herbert Dingle comments, "A Scientist could scarcely care less for what has fascinated philosophers. He does not regard this as something to be explained. He starts with observation and forms concepts as required to express the relations he finds between them..."

My third quotation is the withering criticism of philosophers by another philosopher (mentioned before), only this one is a follower of Bertrand Russell, Sir Karl

What is time?

Popper: "You see, the history of man is a queer thing. It's a history of a succession of attacks of intellectual madness, of all sorts of strange intellectual fashions. I don't need to give many examples of revolts against reason (such as Existentialism), for we know how strongly certain fashions have taken hold, not only just in a comparatively small insular group, but, in large parts of mankind. Russell saw these things in that light, and so do I...In the long history of philosophy there are many more philosophical arguments of which I feel ashamed than philosophical arguments of which I am proud...Yes, I cannot say I am proud of being called a philosopher..."

I have quoted extracts from the works of scientists and philosophers. I believe it will not be difficult to decide which of these thinkers are speaking the truth, especially given the harsh criticism of philosophers by Sir Karl Popper, one of the great thinkers of the 20th century. Traditionally we are told that the purpose of philosophy is to think about the universe and the world we live in. If that thinking exercise is so bad that one of the leaders in the field is ashamed to be called a philosopher, then we have to agree that something is wrong with either philosophy or how we train our philosophers. I think both are misguided: The traditional topics discussed by philosophers are no longer relevant to the world we live in, and how we train them is also archaic. Of course, it is common knowledge that since Einstein and Bertrand Russell some institutions have started to call what they do "The Philosophy of Science." But I am not convinced because when I browse through some philosophical journals I see that they are publishing articles on the same old traditional topics from Plato to Kant, with particular emphasis on the history of the subject without stressing what is right and what is wrong in the ideas of previous philosophers. Yet that is what made Bertrand Russell's History of Western Philosophy particularly valuable. The old philosophies remain footnotes to Plato, by whom there are two worlds: the visible world of physics and the invisible realm science cannot reach, as Professor William Kneal put it; it consists of the world of "mental events which are distinct from anything in the physical world..." So mental events are to be preferred to actual physical events occurring out there and influencing our lives physically. This is anathema to scientists who are already baffled by the quantum theory, which amounts to the analysis of matter down and down to the invisible sub-atomic particles. In fact it does appear that material physics is finished; the physical analysis of matter is at an end. There is no longer any solid matter to analyze as it has been realized that solid matter is composed of invisible matter down to nothingness.

Of course there is a problem with induction, as there are numerous problems with all aspects of life, especially in medicine. What scientists are saying is that they don't know everything; so many of the quandaries in life defy scientific explanation. But they don't allow them to frustrate scientific activity. They rather get on with it, and they are right because they get results. As Bertrand Russell has warned, in physics for instance, we have to obey scientists

What is time?

on pain of death, because whether traditional philosophy approves or not the scientific method can destroy life.

On the contrary, instead of doing everything in accordance with what the traditional philosophers have said or are still telling us to do, Russell was bold and commendably judgmental; he condemned some thinkers; he praised others, almost precisely as Sir Karl Popper has also done, because Popper was one of the 'great' followers of Russell. It must be stressed that the notion that philosophy has to move closer to science has been known for years. Let me illustrate what I mean with a few quotations from Russell's great book. I remind the reader that we are discussing how Russell, guided by logic and history, science and common sense, praised or condemned some philosophers, and I regard that as one of the best things he did. It is extremely important to judge philosophical ideas by the requirements of human welfare and survival. For it is not disputed that philosophy is necessary; the contention is that, as the quotation from Professor Kneal has shown, not many of them respect scientific ideas or the scientific way of looking at the world, ordinary human welfare and what is necessary for human survival.

About John Mill Russell wrote: "John Stuart Mill, in his Utilitarianism, offers an argument which is so fallacious that it is hard to understand how he can have thought it valid." On Aristotle he wrote: "In reading any important philosopher, but most of all in reading Aristotle, it is necessary to study him in two ways: with reference to his predecessors, and with reference to his successors. In the former aspect, Aristotle's merits are enormous. In the latter, his demerits are equally enormous. For his demerits, however, his successors are more responsible than he is." Thus, in spite of his many faults, especially in his Metaphysics, Aristotle got off lightly. But on Rousseau Russell was clear that he invented evil: "He is the father of the romantic movement. Hitler is an outcome of Rousseau, Roosevelt and Churchill, of Locke." To Russell, Nietzsche was also a merchant of evil, and who can blame him? He said: "I dislike Nietzsche because he likes the contemplation of pain, because he erects conceit into duty, because the men whom he most admires are conquerors, whose glory is cleverness in causing men to die." But Hobbes and Bacon came in for some praise. He wrote, "Hobbes (1588-1679) is a philosopher whom it is difficult to classify. He was an empiricist, like Locke, Berkeley, and Hume, but unlike them he was an admirer of mathematical method, not only pure mathematics, but in its applications." "Francis Bacon (1561-1626), although his philosophy is in many ways unsatisfactory, has permanent importance as the founder of modern inductive method and the pioneer in the attempt at logical systematization of scientific procedure." Also he said: "Spinoza (1632-77) is the noblest and most lovable of the great philosophers." On Schopenhauer he wrote, "Historically, two things are important about Schopenhauer: his pessimism and his doctrine that will is superior to knowledge. His pessimism made it possible for men to take to philosophy without having to persuade themselves that all evil can be explained away, and in this way, as an antidote, it was useful. From a scientific point of view, optimism and pessimism are alike objectionable: optimism assumes, or attempts to prove, that the universe exists to please us, and pessimism, that it exists to displease us. Scientifically, there is no evidence that it is concerned with us one way or the other." Lastly, let me conclude this section with his views about Rene Descartes, who else? The man who made French intellectually as great as those of ancient Greece! Of course he is praised by Russell, and rightly so: "Rene Descartes (1596-1650) is usually considered the founder of modern philosophy, and, I think, rightly. He is the first man of high philosophic capacity whose outlook is profoundly affected by the new physics and astronomy. While it is true that he retains much of scholasticism, he does not accept foundations laid by predecessors, but endeavors to construct a complete philosophic edifice de novo. This had not happened since Aristotle, and is a sign of the new self-confidence that resulted from the progress of science." This is a description of a Philosopher/Scientist, not simply a thinker in the mould of Professor William Kneal and his 'mental event' or Professor Wilfred Sellars, who, according to Herbert Dingle, "...reverts wholeheartedly to the Mill type of bypassing of scientific activity."

Science is always good unless it is deliberately misused by some evil men, mostly for political purposes. Otherwise the basic aim of science is human salvation. However, as we have seen, philosophy is not always good, but very, very important because scientists cannot adequately think about the value of what they do; yet there are thousands of them; theories abound; some contradict others. It is necessary that a class of clever persons, or thinkers (suitably trained), makes it as part of its business to look at what scientists do overall and advise them in the human interest. I believe only scholars who philosophize can do that. They do not necessarily have to have doctorates from august universities, but they must think as philosophers not so much through speculations as through what scientists are finding out about people, the world and the universe at large.

Training, or education, makes a man. Lawyers coming out of law schools, as a prime example, always seem to have been brainwashed to be instinctively against crime, democratic and fair in their judgments about human frailty and dead against torture, unless they are basically evil in nature. Otherwise even when a person is found guilty of a heinous crime, his lawyer would plead mitigation and mercy for him or her. They even campaign for improvements in prison conditions. Why can't we train philosopher/Scientists as well---that is, to be equally fanatical about science in its noble pursuit of reliable truths in the physical world, medicine and society itself? To put it another way, why can't we devise a system of training to make competent people, men and women, capable of understanding science and instinctively think about it philosophically. Let me suggest how we could go about this training, on the understanding that I am a fallible human being and that my suggestions may not be the best imaginable. However, somebody has to start the ball rolling.

First, the philosophers: actually nobody can prescribe what must be taught to scholars to make them Philosopher/Scientists forever. The subjects will keep changing. But, on the whole, as we have seen, if philosophers concentrate on the Platonic mental events so that they can only write footnotes to Plato; or, conversely, if they chose the Mill type of bypassing scientific activity, they would have nothing of interest to say to scientists. Yet science dominates every aspect of human life and must be given competent intellectual, or philosophical, guidance. With this proviso, and roughly speaking, I think that for future philosophers to be able to understand science properly, they will have to be taught, among other things, subjects like logic and mathematics, metaphysics and astronomy, ethics and psychology, and also the general principles of scientific medicine. They cannot ignore history and literature, meaning the works of great writers and literary criticism. All the sciences, particularly quantum physics, will have to be taught in philosophical classes. Bertrand Russell's books must be read, and read very well. Methods must be found for teaching Relativity, special and general. It must be taught without the Minkowski contribution; then, conversely, taught with it, so that scholarly will come to understand the 3+1 formula as opposed to the Minkowski purported equation of space with time to abolish the 3+1 system and contrast the differences. By showing students what is meant by merging space with time and why the Minkowski technique for doing so is untenable because of his use of imaginary time coordinates, they will come to understand that reality is based on observation, not on somebody's mathematics alone. For if physical reality is 3+1, mathematics cannot change it to one of four-dimensional continuum. The above tentative suggestions may be found useful in planning the philosophical education of future generations of philosopher/scientists. But they must also learn about ethics, economics and politics, particularly about the dangers of 'Failed States', both external and civil wars, the importance of peace and constitutional law---these subjects show how important thinkers can be. Now let us look at the education of scientists to make them appreciate philosophy and philosophers.

The majority of scientists are woefully ignorant of the other branches of science. Also, it is they who are moaning that they cannot understand relativity without the Minkowski formula, which means that, since the Minkowski formula is logically flawed, they do not properly understand relativity. So the above training programme plus their own specialist fields will be required---not much else needs to be said on this subject. The only problem concerns the complex mathematical interpretations of general relativity based on the concept of "curved-space-time". By this monster time travel is said to be possible. Of course, if space is the same thing as time then it could be. But the fact is, if the Minkowski equation of space to time is false, then when space curves it cannot take time with it. The whole concept will have to be re-examined on the basis of the 3+1 formula because it means four-dimensional space or 4D geometry does not exist, yet scientists are propounding all their theories on the basis of four-dimensional space or space-time. It is a serious problem and I don't know how they are going to solve it, for so many theories since Einstein will have to be reviewed. In plain language, many scientists are going to have to try to understand relativity without the concept of 4-D geometry, or the idea that space and time constitute one entity---a lot of scientific theories are going to be discarded as happened after Einstein when the eather hypothesis was abandoned.

Unashamedly, of course, this chapter is entirely about the status of Einstein as a thinker---and God knows he was some thinker---and the changes in our intellectual traditions that have sprang up after him. But was he also a philosopher, perhaps even one of the greatest of all time? We first must show what makes a philosopher. A philosopher is a professional thinker who has provided an insight about one or many of the aspects of nature and man that could not have been discovered from any other source. If his discovery is great, we call him one of the great philosophers. We know Einstein was a scientists and one of the greatest; but did he qualify by the above definition to be labeled a great philosopher?

Well, every reader can see that I adored him enormously. His greatest scientific achievement is call general relativity; in fact, it is entirely about gravity; and we regard his discovery as one of the greatest in science because it was so strange and yet true. We are all astounded and cannot imagine how he got his knowledge that gravity is caused by the curvature of space---although Riemannian geometry did help him a lot. But, strictly speaking, gravity is part of science. It is physics. It is not philosophy.

The philosophical aspect of his theory of gravity is the insight that the cosmos consists of two metrics: the special relativity frame and the general relativity frame. This was recognized at once as an extremely important original philosophy: two worlds in one universe. One in which life could exist and a second one where there is nowhere anywhere for life to evolve and flourish. So there could be no time there as well. For, as Russell has observed, time is a construction—sentience is required; and you have to have somewhere to live before you could undertake any kind of construction, physical as well as mental!

Without Einstein we could never have known that there is another metric in the universe (called general relativity) where life is virtually impossible. And since this has been confirmed, he becomes a great philosopher by that apparently simple observation or discovery. Yet even that is dwarfed by another momentous insight: the notion that the Lorentz concept of 'local time' is time. Lorentz thought it was a mathematical curiosity, and put it aside. Later he was to admit that he failed to discover special relativity simply because he did not attach due importance to the fact that local time means somebody creating an alternative time as against absolute time, which was the

What is time?

insight Einstein needed to arrive at the idea that time can begin by anybody from anywhere and therefore it cannot be either absolute, fixed, or generally permeating the cosmos as a rigid entity; fixed, as many people believed, by God---all because it is so mysterious.

Thus Einstein more than qualified for the grand title of a great philosopher/scientist for that one idea alone---yet we all know that there was more to come, much more. For the point is, it is not the size of a tome that determines its value; not the complexity of an idea that we worship; not the intricate mathematics of any proposal we want to see. What makes any intellectual contribution important or even momentous is its value in our scheme of knowledge. Some tomes are valued mostly by paper manufacturers, so heavy that they could be lethal missiles in gang warfare. One example is a book published by Cambridge University Press in 1995. A major tome by an incredibly high-brow team of international professors and notable scholars. It was entitled "Time's Arrows Today: Recent physical and philosophical Work on the Direction of Time". Yet a few years later Professor Yourgrau wrote his book about the forgotten legacy of Gödel and Einstein concerning the fact that time cannot exist under relativity. He said he was telling us about the "revolutionary notion of a world without time." So who can we believe? The mighty team of international scholars' thesis about the direction of time (something they know to be definitely in existence), or the legacy of Gödel and Einstein to the effect that time could not exist under relativity? Only a brave philosopher/scientist would dare to hazard an opinion; no one else should dare to intervene. One side or the other is wrong; yet we know that one side is so powerful that if it got angry with any amateurish (non mathematical) intervention, some people could lose their jobs!

As a retired diplomat well over 78, I have no job to lose. So I dare to state that, technically there is no time if each 'body' has to have its own time so that "there are as many times as there are inertial bodies". If any inertial body has not (yet) invented its time, it will of course have no time until it has created one. It means to have time sentience is required. Somebody must be there to invent the time---or place the points for the year and pare it down to the seconds. However we know that, thanks to Einstein, this applies to quantified time only. Otherwise things grow in the absence of sentience; and it takes time for things to grow. Therefore there is natural time---natural periods during which things, events, action and growth occur. What we add, the human contribution, is the process of reducing it to culturally manageable units, using repetitive cycles. Given these facts, the mighty team of international thinkers failed because they knew nothing of 'quantified time'. They just talk about time; and Professor Yourgrau is right, for time in that ancient sense linked to motion does not seem to exist, since there is not only one form of motion but billions, and they do not all pass by in tandem---i.e. there can be no general passage of existence because existence is multitudinous, and we all, billions of us, see the world differently. Only deliberately 'constructed' or 'Quantified' time (human in origin) unites us all.

On the other hand, there is time in society. Wherever there are 'Beings' there will be time, for man cannot live successively without time for long. What to do during the day, and what to do at night require time for regulation; similarly, what to do during the coming days, weeks, months and year, requires time for planning. Hence quantified time, being an insight born of the combination of scientific facts and philosophic intuition. We need many more of such thinkers especially as society grows more and more complex; and while training cannot produce anybody half as good as an original one at least we can try.

Chapter Eleven: The Status of Earth Time In The Universe

First, the definition of time so that we know what we're talking about. It is not a time system that runs through the whole universe, for there is no longer a universal time. Perhaps it will do no harm to quote Russell on this again: "There is no longer a universal time which can be applied without ambiguity to any part of the universe; there are only the various 'proper' times of the various bodies in the universe." The only time we know of is earth time, which is based on the orbits of the sun and is

What is time?

strictly determinate. The year has to be repeated for our time to continue, otherwise there is only one year. That plainly means we have only discrete time of limited application, but the cosmos has no time at all, and what appears to be time in interstellar space is caused by either chemistry or accidents, which we may generally refer as the natural state of existence in the inanimate world, where brute force is king.

Here on earth the whole idea of time---as complicated and mysterious as it is---is simply a mechanism for showing 'how long it takes' or 'how long it lasts', but of what? It surely must be 'contact'. Again time is reality as perceived even as it is changing continuously at the atomic level, such that a digital camera can capture several images in an instant---all different at the atomic level. The actual images of reality are converted to 'units of time' by the repetitive cycles we use to reckon time. So while the time itself does not move, the movements of the cycles create the illusion of the march of time. Time's actual movement is achieved through replication of its units---the years, for instance, replicate to become the centuries; it is the same with all the units of time which are fractions of the year. The 'March of time' concept is extended to history, reckoned from the last or past minutes. Yet history is rather the march of events not time. As the mere apprehension of reality, time does not march at all. It is contact with reality as it is. Whether your contact (or what you perceive) will last a minute or hours depends on the cycles used for reckoning your time. We on earth all have the same units of time because we use one cycle (the year) to reckon our time.

Without contact there is no need to know anything because there will be no evidence of anything's existence other than oneself as a solitary 'Being' (with no need for time reckoning.) So, given that it is man against the rest of nature, contact implies everything external other than ourselves as individuals. Even then touching parts of one's body is also contact for the reckoning of time expended. This or these contacts begin from the womb; hence the sense of time is intrinsically part of our nature. This may be the reason time is inseparable from life and extremely difficult to explain as 'constructed' by man rather than divinely bestowed from the heavens. Yet scientific or logical thought has absolutely no truck with mysticism. Either we know something or we don't; there is no need to introduce revealed knowledge into discussions of material reality. Of course, I am aware that many scholars believe that mysticism has a role in scientific thought, but Bertrand Russell thought otherwise, and therefore did not mention Wittgenstein even once in his great book on Western philosophy, because his work amounts to nothing but logical mysticism which aims to put an end to physics. This is as close as calling him insane.

After the definition of time (without mysticism, revealed knowledge or fantasy), let us begin this section of the book with one important statement of fact about time: time consists of units---or single moments---as Professor Whitehead prefers to put it. The cardinal example is the earth-year. However it is utterly impossible to define any unit of time in the abstract such that it can be recognized when conditions for it are fulfilled. In other words, it is impossible to know what conditions will produce (exactly) what unit of time in any part of the universe. The year itself can never be defined in logic. Therefore what a second of time is can never be defined in such a manner that people can recognize a second when they see it. For this reason alone earth time cannot be valid in any other frame, including that of general relativity, and I think cosmologists need to be reminded of that. Now let us consider the status of earth time in the universe, knowing that it cannot be applied anywhere else, being a product of the earth's peculiar postulates and periodicities. I have come to the conclusion that it is possible our time is created with a unique set of circumstances, and that on other planets it may not be the same or even called 'time'.

Now, how life appeared on this planet, and for what purpose if any, as a composite question, most dire, is the only conundrum greater than how we get the time by which man lives his life, and how the time moves on, or continues perpetually. It is not surprising that before the rise of science (and even still now) many thinkers believed that it is time that gives us the right to live; the general idea was that life is based on time; whereas the scientific study of time indicates that time is based on life---somebody must be there to count the orbits of the sun as years (consisting of the seconds and all the rest of it), or there will be no years and no seconds, only bland existence in a senseless world. Time is the motive force in the mind. Every movement, thought, action and inaction occurs in time. Time enables us to act and be human since every action takes place in time. So if time is created by man then it must be two in kind: the internal sense of time and the external parameters by which we measure intervals. The sense of duration (the internal sense of time) is natural; what we create to link it to physical reality is the external time, and is strictly based on points as applied to space. Is this man-made time any use in the cosmos at large? I think not, and will discuss it in a moment. This brings to the fore for a brief discussion the question of whether or not there is time (or there will be time) in the absence of somebody counting the orbits of the sun as years. As a question for serious debate it is akin to the problem of the existence of God, because it is so mysterious and of great philosophical significance. Is there time in the absence of human intelligence? I think not. Without the human mind what appears to be time is rather the natural motivation of things caused by chemistry or accidents, sometimes called 'law of the jungle'---almost all of them caused by chemistry, brute force and accidents..

My personal view is this: there is always motion of the kind we associate with time, like things moving on, growth, decay and ageing, even leaves on trees waving in the air. They are also action and they take time. Some thinkers assume this process to be time moving on, perhaps even silently, without the intervention of human intelligence, or rather not even requiring man and his mind at all; that,

What is time?

naturally, trees will take time to grow, for instance. As a matter of fact, such natural events do not constitute time per se; instead, I see them as events occurring to certain objects in their own worlds, or in their own 'Beings': a tree grows, a river flows, a person is ageing, and so forth. It is not time we can mechanize in a clock. Things live their own lives to which time, once mechanized in a clock, can be applied; but the passing of their growth or decay is not quintessential time. It is not constructive or quantified time. They are not in philosophical dispute.

Although the general growth of a tree can be explained as "time going". It is not the kind of repetitive cycles we can mechanize in a clock. All things that move or grow can be set to mechanized time; whatever happens to them happens through the passage of time. But it is not correct to call the invisible growth of the human hair as time. It is not time but the natural chemistry of the human hair. Rather it is correct to regard it as "time going"---invisible time going. To know how much time in physical and visible reality you have got to rely on mechanized time based on regular or repetitive cycles and count them as the rates of time, the years, for instance. If we regard every growth and motion, backwards and forwards, as time we would end up virtually in a confused world of myriad of time systems. So for scientific and logical thought we rely on mechanized time; all references to time should be reserved for mechanized time suitable for universal application in one inertial frame. (See 'Time and Quantified Time' in Appendix I below.)

Again, if we consider any motion as 'time' rather than as 'time going', the implication will be a reference to pre-existing universal time, not one created locally for local purposes as space-time in specific units which can only advance 'unit-by-unit'. On the other hand, the idea of taking any motion as 'time going' (in units), means the time must have been established already so as to have it in specific units. All references to the concept of space-time are to be understood as time obtained from space with points, not in the sense that time and space constitute one entity by the Minkowski mathematics---the man who succeeded in fooling the rest of mankind with mathematics, He's not a fraud except that mathematicians can gullible, even childlike. Another one got up to claim he'd made a discovery that time will end in a black hole, when the time had already been proved to be discrete, no longer universal and utterly impossible to travel through the universe to end up in a black hole unless it's carried by mathematicians on their bald heads.

It has happened in the past that many things were used to mark time: the shadows of trees, of mountains, of houses, even human shadows. But since the earth-year is now used for time over the whole planet (albeit with zonal variations), we tend to interpret motion and events in terms of the amount, and length, of units of time expended, being expended, or, futuristically, to be expended. It is the same under the relativity notion of time. Time is now known as 'space-time', being the product of points as applied to space; space-time is necessarily discrete; it does not run through all nature. As such we have to search for the method used to establish our time as space-time, which, of course, is (and has to be) limited in its effects to this planet alone. In this situation, the only way to have one time for any inertial frame overall is to mechanize some repetitive cycles (the year, for instance) as time for all and sundry---with the inevitable zonal variations, depending on the size of the frame or planet. The clear philosophical implication in epistemology is that time does not exist in nature at all, if time is defined as "time in the clock". Yet only time in the clock is scientifically relevant. Sentience is required; intelligence is necessary; the ability to count is indispensable; and a theory of numbers is absolutely essential, all of which makes it seem human in origin, but based on the natural sense of time as "duration" felt in the mind. It should be noted however that duration also requires points, therefore contact is still required or implied as the origin of the sense of duration and time.

Even the concept known as "the passage of existence" is meaningful only as "passing through a human mind". Somebody must be there to count the orbits of the sun as years, and have the intelligence to sub-divide the year down to the seconds, or there will be no years and no seconds. Thus, precisely like language, time arose out of necessity. The multifarious activities, motions, and events in existence (growth, decay, to and from, up and down, backwards and forwards, and so forth) all occur to individual things and beings---human and animal, rock and plant. They are individual occurrences; things living out their own lives through their physiology and chemistry, physical and organic. Otherwise there is no time as constructive moments in succession

Let me stress that, nowadays, we suppose that what happens in nature, like growth, the flow of rivers, and so forth, are seen as the process whereby objects and beings live their natural lives. Each and every one can be set to time---but where is the time to set them by, since we do not accept that time is just there? They can be set to time only after an acceptable concept of time has been established with intelligence as applied to some repetitive cycles in conjunction with the sense of duration in the mind---in short, only after quantified time has been mechanized in a clock.

Thus one of the consequences of the space-time idea is that only mechanized time is true time for general application; all other semblances of time are just the chemical processes of things; but they are not useless in the scientific study of time because they can be set to time. The irony is that the attempt has been so successful that sometimes we tend to believe that time is naturally in existence, and say, for instance, that the flow of a river is an example of time going; yes, but by whose time? Without time in the clock, the situation will be confusing, for nobody would know how much time is going. In a way the passage of existence and ageing is time going; but you have got to have the time already in a clock to know how it is going, and by how much.

What is time?

The huge variety of objects in existence, each living its own life according to its chemical make-up, means that, although the growth of things may be seen as "time going", but that is not time for general use. I reserve the word 'time' for time in a clock, written in mathematics as ct. What that means is that it is time for an inertial frame, according to Einstein's theory of relativity. It means time requires points and mathematics for linking the internal sense of duration to external cycles that occur repetitively. Such a time is the creation of the human mind.

And here is the crucial question: following from the above, another puzzle arise, namely, can we apply our parochial time to events in the cosmos at large, say, regarding its past, present and future of bodies? How can we suppose that going round our tiny sun and calling it one year can be used as the yardstick to tell the age of the whole gigantic universe? Ten orbits is ten years but the period is so short because our sun is puny

And while we are discussing such matters, what about the definition of the time content of one year---or how long is one year, and how do we measure the length of one year in temporal terms? If we use distance it is the as saying we can only know how time is passing and never what it is. For instance, as a matter of concern to all of us on this planet, when one year passes, you know you are aged one more year, that one year of your life is gone with it, but how much of your life span is gone---how do you measure that?

Thus, in my opinion, it is not very helpful asking how old is the universe, but 'How old is the universe by our time?' There is no other yardstick. On realizing that the largest (or longest) unit of our time is the earth-year, which is just a measure of one orbit of the sun, estimating the age of the entire vast, gigantic and mysterious universe by this yardstick ceases to have any credible meaning.

Again, how at all does the cosmos figure in all this---that is, in the nature of our time? We believe we are here on our own as far as life is concern. The being of things offers no comradeship because everything is absolutely individual, except, perhaps, things like the branches of trees, where one thing depends on another. A universal time might give the impression that somebody is in charge and knows of us because he has given us part of general time, our version of it. But once our time is seen as uniquely our own (and completely secular) its range becomes doubtful when applied to the cosmos at large.

As is common knowledge, there are stars so immense that a million of our petty sun can find room in them. Given this fact, does anybody really believe that going round our tiny sun and calling it one year means we could determine the ages of events in the universe with the mere arithmetical accumulation of single units of our relatively short year? The abusive lunatics on the internet were wrong to attack me with personal insults; for I am only asking questions. There is no pretence that I can have answers to them. A thinker writes to invite discussion. Nobody who is not insane will write to dictate to the rest of mankind. Like the rest of us, I am also groping in the dark about everything. But we are now coming to the view that

What is time?

certain questions, since Einstein, undermine religion because the latter is not logically based. It is not the thinker's fault that it is not logically sound. However, it means we've been misled during many centuries of religious thought and obedience.

Even worse, apart from events, can we really seriously use these short years to determine the actual age of the universe itself? I have already said it is not correct to ask the question 'how old is the universe?' The proper question should be 'how old is the universe by our time?' Otherwise, by whose time, since age is related to time? The longest unit of our time is the earth-year, fifteen billions of which are mere fifteen billion orbits of the sun. We can think of acceleration that would give these 15 billion years in a short time! In any case what about the time before the sun came to be in existence? Furthermore, what of the time before the earth formed from interstellar debris into a planet and began to circle regularly round the sun, each of which is one year to us? Even this is not accurate enough. We should begin from when mankind acquired the facility to count the orbits of the sun as years, a most recent event by all accounts.

The religions have a lot to answer for. As infants or imbeciles, we are forced to worship anything in the name of God. It is coercion, and it is criminal. However, I am really surprised that the way and manner we get our years up to the centuries and even the millennia (merely by counting the passing years), has not undermined the so-called serious scientific theories of the age of the universe. These serious thinkers tell us that the age of the universe is between 13-15 billion years. Fifteen billion years for the age of a universe containing stars so immense that millions of our petty sun can find room in them; and especially when, in fact, one year is just the time for going round this tiny sun? I believe that this period is not even long enough for the formation of galaxies with their hundreds of billions of huge stars.

Here is another puzzle: how long is one year is a question I posed earlier. There is no answer and I know it. Let us be absolutely clear about this, we can never tell the length of the natural time period that one year contains, and yet the years are just repetitions of one year. Also we cannot use parts of the divisions of the earth-year (say, the months, days, or weeks) to define the temporal length of the year in cosmic terms. So we do not know how long in temporal estimates is 15 billion years other than that it is 15 billion orbits of the sun, yet this sun is, on cosmic scales, too small to require much time to circle it.

Frankly, the questions are endless. The age of the universe, in my opinion, is better left alone. It is not suitable for serious study, and it is not important anyway. The truth of the matter is that these studies cannot be justified on any grounds whatsoever. The objects are simply too far away to have any effects on our lives, the only sane reason for studying the cosmos in cosmology apart from vanity, is intellectual satisfaction. But astronomy is different; it is the subject we need to study very, very seriously.

What is time?

We can legitimately study stellar events without worrying about the actual age of the universe; so, let us just say it is not amenable to human ageing concepts. I doubt that any inspiring theories will be missed by forgetting about the age of the universe. The universe is unimaginably vast; personally I shudder to think of its extent, nature, and how it came to be; these are human terms; they don't seem to apply to the cosmos. The mind boggles. And that is because we have forgotten the lesson from the nature of the quantum, which is that the terms we employ in our discussions of the universe and the sub-atomic world are human terms unknown in the universe outside a human mind or head. Some particles can appear in two places 'at the same time'. Yes, because they do not know what is 'the same time'. The universe does not respond to our time, concepts, categories or the things that pass through our minds.

Because philosophers are not properly appreciated by scientists, many momentous novelties introduced by Einstein (the reason they called him 'Philosopher/Scientist') are not given the necessary intellectual attention. Wittgenstein had much to do with that. At the time A.J. Ayer and Russell were stressing the importance of logic and philosophy, along came Wittgenstein to preach his logical mysticism to annoy scientists. Let me reveal a little of what has been missed by not paying attention to the rational thinkers whose only fault was failing to preach what people want to hear.

Apart from our knowledge of bulky matter (or normal perception at the eye and sensory levels), there is a hidden reality we cannot access at all, or can only partially reveal through mathematics and logic, even then in such theoretical formulas that very few of us are capable of following. After all the world has not been prepare to receive human beings in comfort; it would be different if it were so arranged. That hidden world is alien to us as we, also, with all our 'different-worldly-concepts' are unknown, even strange, there. These are different levels of existence. A simple illustration will make the idea clear: the golf and table tennis balls may look alike from a distance, especially if they have been deliberately equated in size. Only by examination can the differences between them be revealed. This trick of the eye is used a great deal in the Movie Industry, especially by Directors and Producers. The crooks also use it to deceive. But in theoretical physics mathematics can be used to sort things out. Yet the worst oddities still remain; or, to put it another way, only a few of the oddities have so far been revealed. Much of nature is alien to us as we are to it. There are millions of things out there we don't know, some of which startle us from time to time. We first noticed this mystery during the development of the quantum theory and should have learnt the lesson well. There are things we cannot even learn until we are capable to learn them due to lack or cerebral competence or capacity, which may sound like a paradox, though it is truly the natural course of events in the cosmos, since we are always growing in brain capacity: an infant cannot learn to drive or fly an aircraft; the computer and internet were not discovered until recently, even though the facts, technical details and materials were there hidden in nature and waiting for discovery. It is not impossible that in future human beings can be compressed into a chip and transported to distant planets to be reassembled and live normally afterwards to colonize new worlds---especially those whose surface gravels are all diamonds, if we're living under a plutocracy! It may be one way of moving to live on other planets, but at the moment we can't do so; and while the idea will be branded fantasy today, if and when it happens people will not even remember that it was once fantasy; they'll rather say it's the normal course of developments or progress.

We have to remember that a human being with all his massive brains, knowledge, and theories about the cosmos is physically less than a tiny drop of water in the Pacific Ocean by comparison to the universe, even to a galaxy of mere hundred billion stars. So far we have failed to make sense of it anyway; and it is about time to let it be. Astronomers can continue to skirt the fringes of some stars; yet again the nearest star is several light years away. I know of ingenious theories of rapid travel across space, but do we suppose that the human body can bear the strain of these velocities even if they were feasible---and for what purpose?

Humanity should always be understood as limited to the earth and its dwellers; as annoying as it may sometimes seem to be, your humanity and love is to another human being who can appreciate them, and there is nothing sweeter and morally inspiring than the appreciation of your fellow human beings. That is why many nations have 'Honour Systems'. It is about time we thought more about the world than the cosmos; there is nothing it can do for us whether we are rude or humble to it. It is a cruel world; no efforts should be spared in trying to minimize its harshness. That is the first lesson in humanity.

Time only became a major problem in science because of Einstein. As Professor Sir Arthur Eddington has remarked, it was not so before his researches about time. The scientific problem of time is different from that faced by philosophers. We have a situation where we are using time daily but cannot define what it is. Yet Einstein made time a separate co-ordinate in the determination of physical reality. That's what matters. The units of time remain the same; only the metaphysics of how we get them has changed.

The basic unit of time, the year, is virtually indefinable; while other units of time can be defined only in relation to the year as sub-units thereof. At the same time there are elementary conceptual difficulties in science about time: on the one hand, time is an artificial concept, called Space-Time, a 4-Dimensional metric of whose existence we have absolutely no evidence except that of imaginary 'thinkability' (as used by Einstein himself), or the Minkowski mathematics based on the square root of minus one ($\sqrt{-1}$.ct...) for time. This is ruled out as untenable in logic---if i is representing time as stated above by Einstein, no less, then

What is time?

what is the ct in the same equation representing? It certainly looks like an elementary logical error.

And yet, on the other hand, in all science we learn that time is naturally in existence, and always passing. But what is it that is always passing, if not the regular or repetitive cycles we use to reckon time? The cycles we use to reckon time of course go exactly with time, thus misleading us into thinking that the motions of the cycles are those of time itself.

We know they are not because time is discrete. It is based on the year and so it is bound to be discrete as the year is determinate, and passes by as the 'years' through replication not as a stream. Discrete time cannot run continuously like the cycles. So what is time? What does the actual physical nature of it look like? The natural aspect of time is the sense of duration in the mind. That is incontrovertible. Duration as time in the mind is always there, related to the sense of things enduring as part of the memory mechanism; for it takes time 'to endure'.

Some scientific thinkers have settled for a definition of time that equates time to being; that time is identical with 'Being' exactly, and that ageing is time going and taking life with it, not the other way round. The problem with that definition is the concept of points being required to apply to space to get the time---that is, create the time in the first place, once we agree with Russell that we construct our time under relativity. For how otherwise can we divide 'Being' to get time in units? We have to use repetitive cycles or motions. Also the idea of dividing time implies that it exists in some kind of a thread running through, yet the year upon which everything about time is based (related to or as fraction thereof), does not run through nature. It is one orbit---only one orbit of the sun and ends there. Our time is ended after one orbit of the sun. Another orbit is another year, not related to, or dependent in any way upon, the last year in the sense that it could not exist without the past year. Thus the idea of time running with 'all being' (as a universal time) has now been abandoned. That is another problem of time solved. But it is interesting that all this goes back to the Russellian query after Einstein's demolition of absolute time: "If cosmic time is abandoned, what really is measured by the clock"---in other words, how are we going to define time without the cosmic aspect of it? That is the question I have been trying to answer in this book. And I can reveal that it has engaged my attention for the past fifty years, enduring mockery, abuses and snobbery from the high and mighty in the academe and elsewhere! And now writers and thinkers have social media abuses too to contend with. If it gets worse we'll simply have to give up.

Appendix I: Time and Quantified Time or the Passage of Time

We are all fond of using the word 'time' loosely to refer to the passage of existence in any form whatsoever. That may be called 'the unscientific' notion of time. In logic, science and philosophy, however, time is what Professor Richard Feynman called 'how long we wait'. This translates into the concept of 'how much time', or quantified time, so as to be able to tell how long we wait in mathematical language for universal application.

In any serious discussion of time, it does not make sense to just mention time. It may take centuries to understand that 'Being' on its own is not time; it's true that you have to 'be' in existence to know how to apply points to create intervals of time between points, but being on its own is not time; being has got to do something to nature (that is know how to divide it into periodic intervals) to create time---that requires sentience. I insist that such a time system cannot move; what appears deceptively as the running of time consists of merely the motions of the cycles used for time.

As discussed above, motion is not time either. It shows time going, but the time will have been created elsewhere beforehand. At best motion is silent time; but I think all that is chemistry, for chemical processing can impose (or require) a period of waiting at the visual level, which is the same thing as time. The truth is that all sorts of things may be called time but none can show how it began. We can only rely on how we experience it, and that is through the use of points as applied to space to create time intervals. Logically, this is the best we can do either in science or logic. Everything else is sheer religious humbug.

So motion is not and cannot create the time we can mechanize into the clock; the simple reason is that it is multitudinous. But we can count cyclical motions and call each cycle say, 'a year'. If this cycle is continuous we can have years and years in perpetuity; and a year of course is time; it allows us twelve months to do whatever we want to do; it is also the standard measure of age.

And that, precisely, is what we do to get the time to program into the clock. However this can only show how much time is passing by means of our physical cycles and never the real thing. We count mere physical cycles and call them the rate of the passage of time---but what is the true

What is time?

nature of time? In my opinion time is the same thing we call life. The secret of life is the same as the secret of what we call time, with the proviso that all we can ever know of this time is how it is passing by; therefore the implication is that we can never discover the secrets of life either. Obviously we have to be in existence; and we are in existence (I think therefore I am!). So it means we are living (we're there or here) and choose to use a regular motion as the rate of the passage of time. The cycle does not call itself time. It's mankind that regards it as the rate of the passage of time, or the measure of duration to guide his actions. That's all there is of time. It is not as scary as was previously thought; it's merely a device to guide our actions. One can even tap the finger to the same effect. Let's say a thousand taps means an egg is cooked or done; that's not different from saying ten cycles (minutes) means an egg is cooked. The earth's orbit is so long that we've had to sub-divide it into smaller units of time. But every second is part of the yearly cycle, and therefore logically part of a cycle. A second to go is not yet a complete year. But deciding the when by means of the logical study of time brings in notions of earth time.

The nearest we can get to the definition of life is the logical definition of time, for the two are closely associated and don't seem to be separable. After many years of thinking it all came to me one day encapsulated into one word "When?" When is anything? We can't have any existence without when (the time) it is or was in existence.

Thus life is time and time is life, since we cannot define life without the "when?" it was or is there.

Ironically the logical definition of time reveals it as merely how it is passing by through the use of physical cycles. Thus a great conundrum, juggled round and round in the most exhaustive manner, becomes the greatest mystery. It is that time is life and life is time, simply because every second of existence is time; once you're alive you're expending time. It's not the same as saying it is the time allowed by God. It's slightly different; though I concede that the religions came close, very close. They've always had some back-room chaps bearing their intellectual burdens and some of them, like Johannes Kepler, were very good indeed. The difference between my theory and that of the religions is that I am asserting that time is life and life is time; on the other hand the religions claim that life spends the time already allotted by God. Also, while they have no proof of their views, I have logic and Einstein on my side---with Bertrand Russell as a providential bonus! Time is life and life is time because you cannot define anybody's existence without when he or she was in existence; the quandary is that this time is one of our own creation, or construction. So, again, we have to theorize on the basis that man has a hand in the logical definition of existence. Thus the fields where man the observer cannot deal direct with reality are now three: Plato's simile-of-the-cave, the Einstein 3+1 formula, and our present definition of what we mean by existence---that nobody can exist without his when (time) of being there, but we construct this when ourselves out of the parameters we find in our environment. By the way, the Einstein 3+1 formula is included because of the time element---and we create the time---so it means we contribute to the nature of physical reality as perceived. That is to say, we determine physical reality from the three aspects of space and matter plus time, our time. This is the reason Minkowski incorporated the time element in his equation so that we can write $S=CT$ to represent all physical reality and it is for the same reason of mathematical economy that mathematicians insist he is right. Unfortunately intention (or human desire) and physical reality are poles apart.

Again, my theory means your very existence is convertible to time the moment you are born not that the time is what is permitting your life to endure. The when of your existence comes in as soon as you're born and never stops---time is unavoidably continuous. Once a person is alive time takes over the control of his or her life in the following manner: you're alive. To continue to live on you've got to live strictly in accordance with the earth's motions and environmental conditions. These are what have been converted to time, based on the earth's motions. This time is unavoidably continuous because the earth never stands still. Thus from birth a person is controlled by the motions of the earth as we have converted them to time units---therefore time controls life. To be is to be part of the yearly cycle or be spending part of it---you cannot be without spending time. Nobody can exist without the 'when' or time of his or her existence. Life and time are inseparable. So to establish that time is secular is, for me, the greatest philosophical intuition or insight. I think it solves the last conundrum about time and life too.

Deciding 'how much time' by means of regular cycles is the main job of the interpreters of time. The context of any proposition (in science, mathematics and philosophy) must always show or imply the sense of 'how much time' in it, or expressly show the quantity of time proposed. Of course, time may pass when one is not conscious of it. But in all cases, when one wants to know how much time has passed, or will pass (as in futuristic propositions), mathematics must be used to quantify the time. And let me stress again that we quantify time by the use of external cycles in union with any sense of duration of anything whatsoever.

Quantified time is 'time in a clock', any clock at all. And the clock, any clock, can only show time as independent of space. Space-time is automatically quantified as it is derived from space with points, which is the only reason for calling it 'space-time'. Discrete time can only pass through the succession of the individual units. On this point, Leibniz was absolutely right when he said time is succession. What was lacking in his day was the concept of discrete time; with this new concept in our post-relativity world, we can now see clearly as to how time passes and seem continuous through the succession of its separate and individual units: second, second, second. Plus the hours, weeks and months all the way to the year, which also passes in the form of year after year after year.

What is time?

It may seem surprising, the springs of a thousand legends, giving rise to supernatural speculations, that we have an extremely ingenuously smooth time system, so cleverly structured that it is there when we are born and there as we die, and always passing by. For this reason we know that "Time does not wait for anybody". Scrutinized under a logical gaze, however, time is not so rosy; it is only one moment, repeated to pass by and seem continuous so that arithmetic can be applied to its accumulations. This, as we know well, happens when we reckon time for futuristic planning, and backwards as in historical narratives.

But for the union between the sense of duration and external cycles giving us units of time out of the moments of time, time for the clock would not exist at all. Presently philosophers see time as rather a straightforward pragmatic entity, albeit not as simple as it is normally supposed. It is partly a confidence trick that makes the clock work continuously, the trick of continuity is in the repetitions of the seconds, or of the units of time, all of which are to be understood as single moments---the realities---of quantified time. It is also partly physical (using physical cycles for the process of quantification); and partly philosophical, i.e. according to Einstein without time physical reality is indecipherable, or cannot be properly (accurately) determined, hence his equation for motion consists of the three spatial coordinates plus time in the 3+1 formula of physical reality. This, of course, is contrary to the Minkowski formula and I think this is much more scientific. It is true there is an element of subjectivity in it because the time is man-made, a human concept 'constructed' by man. But at least it is not as arbitrary as the Minkowski imaginary time coordinate.

To sum up, we have to recall that Einstein made man the observer part of the observed. Plato also made man part of the observed with his simile-of-the-cave notion of perception, meaning we perceive the external world not as it really is but just how we are made to see it. When it comes to time (as the most important aspect of life) the situation is the same. We see the world and time not as they are but just how we are put together by the human architect to see it. Logic, of course, is our principal instrument of perception, theory and knowledge. Thus Bertrand Russell, as the great logician he was, summed the Einstein theory of time up and concluded that cosmic time should be abandoned since it cannot logically account for the nature of time as discovered in experiments. Professor Eddington also concluded that those expressing doubts about the Einstein theory of time were making meaningless noises. The founder of astrophysics was convinced by the secular theory of time. One reason, as I have pointed out, is that the religious notion of time and the nature of time discovered in logic are pretty similar: we don't know what it is but all of us accept and live by the yearly cycle as the passage of time, and have been doing so for centuries---centuries which are just the number of times the earth had circled the sun. Any good logician would sense that the true nature of time was not far away, especially after Russell

What is time?

asked the most important question about time---if cosmic time is abandoned, then what is measured by the clock?

Unfortunately, researchers did not follow this logical trend to try and discover the true nature of time, but rather jumped on the Minkowski bandwagon to propose concepts of time travel as 'a scientific possibility'. It is a sad reflection on the mentality of some writers that they should seek to twist the mind of mankind to concentrate on The Afterlife rather than the actual physical reality influencing human life now.

I've always felt that if this had not happened (with numerous books about time travel selling millions while contrary suggestions are rejected), man could have done really good researches about the nature of time. Minkowski and Kurt Gödel bear the blame. But that is not all. Man is basically more interested in life after death than anything else. Well, if the Minkowski formula for equating space to time is not logically valid, it means it cannot happen, and if so then travelling by space-time is not feasible. We have to go back and research time as a secular entity that is separate from space exactly as Einstein made it in his special theory of relativity. At that stage the Russellian question comes up again---what is measured by the clock? Let us consider this question in the next chapter.

But I must stress that discussing the passage of time is necessary only to accord with popular ideas of time; otherwise I don't think time is ever in motion. To me the reality is that the cycles we use to reckon time make us think that time is running all through the cosmos. Yet time consists of separate moments, no matter how long each moment happens to be; so it can only advance through the replication of the units. Our time is based on the earth year which is so long that we've had to sub-divide it down to the seconds---but the main unit and its fractions advance by replication, not by running. That idea is a mental deception.

WHAT IS MEASURED BY THE CLOCK?

Our time is based on the repetitive orbits of the sun by the earth, and evidently the earth never stands still. If ever it does stop going round the sun, our time system will be completely nullified; but, of course, life will go on. It is inconceivable that all life will be extinguished instantly the moment our time is (mathematically) nullified in that only quantified time would be lost. This is the best proof there is that life is not based on "time allowed", as the religions believe; rather time is a union between the sense of duration and external cycles---therefore man had something to do with the time we have in the clock, the only reliable time, as quantified time.

All the religions speak of "time allowed" for the duration of a man's life. They had to, because the nature of time is easier to explain as a providential bounty than anything else. To be honest, without a cosmic explanation for time, what is time; to put the question in another form, what is the origin and essential nature of time? Everybody believed that it's divine until Einstein and Lorentz found that it can begin from anywhere. Of course, it is assumed

What is time?

that the clock measures time. Even Bertrand Russell talked about measuring time, asking what is measured by the clock---but from where? And what is it that the clock measures? The clock maker will say he invented the clock to reckon time in the sense that everybody knows---but what is that sense of time? The mechanics (or clockmakers) used the day and night system, the moon's phases and other astronomical features of the world as far as they're concerned, just to help us measure our version of a universal time. In other word, they're just as ignorant of the true nature of time as everybody else.

When it is postulated that general time permeating the whole cosmos (and therefore the same everywhere) does not exist, the first implication is that every 'body' (or planet) has to have its own time; it is not coming from the cosmos therefore it must have originated on this planet. So let's find out how it all began. That is the first implication. The second is that, as a result, cosmic time is abolished---although it sounds tautological, it still has to be emphasized, as well, and most clearly because the 'cosmic time instinct' is permanently ingrained in the human mind. One reason is that time cannot be suspended; but the more cogent reason is sheer intellectual incompetence plus fear of the unknown. We are always using it, and so it does not make sense to just say that it is not there. But if it is there, and did not come from the cosmos, how did it begin? And the obvious fact is that it is always there. Even before we are born, and also as we die to leave it behind. Yet it cannot be supposed that each body's time is a version of something 'naturally existing', whether it permeates the whole cosmos or not, with the necessary but illogical (little 'academic') proviso that it may not be the same everywhere but varies with individual bodies in accordance with unknown natural laws.

It is plainly evident that this erroneous sense of time dominates scientific thought. Hence time is not defined in physics; and as a result, the Minkowski fiction makes sense to some scientists. They just say "as time goes by". Only Professor Arthur Eddington has redeemed physics by warning that it must never be forgotten that the 4-D geometry formula is "fictitious and arbitrary"---but they have chosen to ignore him, partly because Eddington was afraid to mention Minkowski by name, or maybe he's just cunning. The era was incredibly sensitive: There was Einstein, Planck, Russell, Whitehead, and the mass of aggressive no-nonsense mathematicians who regarded Minkowski as the genius who made relativity accessible to scientists. Thus Eddington had good reason to be cautious---nevertheless, since then everybody refers to the concept of space-time as 'artificial'. Let me explain another small point about original ideas. Even the originators do not stick their heads on them, because they're never absolutely certain that contrary ideas would not emerge; and we all know that most of the time they've emerged to shame cocky theorists. So even Eddington might have been a wee bit afraid of the pure mathematicians---even Newton was, and if David Hilbert is to be believed, then Einstein too was!

Thus Russell's query is important, namely, "If cosmic time is abandoned, what is really measured by a clock...?" My answer, of course, is that outside the union between the sense of duration and its conversion to external cycles, time does not exist to be measured. The very act of 'measuring' is the time in essence---like moving from point to another point, time is going, so that time becomes 'relation between points', or intervals between points. The cycles are time units (the years, for instance), and the time units constitute the time: a year is a cycle, but it is our time, the basic unit out of which all other units are derived.

However, the cycles are the creation of man for the sole purpose of converting the sense of duration (of anything or any event, like the period it will take to reach the village from the farm before nightfall to avoid predators), to his time units to guide his activities. So the clock does not measure time; it rather reproduces units of time programmed into it by the clock-makers. It should be remembered that the seconds are put there by the clockmaker; but where do they come from? The answer is that they come from the subdivisions of the year. Otherwise the time does not exist anywhere to be measured---the units constitute the time. Without the year there will be no seconds, and the like, all of which are derived as subdivisions of the year. As hinted above, you can even dispense with the year and its subdivisions and tap your finger, if you will not get tired. A million taps means it is time to go to bed, and so forth; outside the units of time, time does not exist to be measured; but the units are the creations of man as quantified time to record the passage of existence in his experience in manageable units for cultural purposes.

I conclude that what we call time is the mere physical manifestations of it that we use (as periods of waiting) to organize our lives. These manifestations (or physical cycles) that we call 'time' are in motion of course: we count the earth's orbits---caused by motion---as 'years'. That process has been taken as the march of time. But we don't know what time is to tell whether it is marching or not. My feeling is that events do march and they have time associated with them thus misleading us into thinking that it is the march of time. In fact we can never know what it is, if it exists at all. The causes of the cycles we call time units may be chemical, inertia, dark matter, momentum, motion, kinetics, delayed-reactions and so forth. They cause what we experience as time. If real time does exist we cannot know it because it is shielded by the parameters we use for time. This echoes the Platonic Simile-of-the-cave again---it seems man just cannot perceive real reality. To me that's not so strange, for we are so insignificant anyway. The real surprise is the incisive power of our brains.

Appendix II: The Principle of Mathematical Equivalence

In nature there is reality and our perception of it. I subscribe to the Platonic simile-of-the-cave theory of perception. In the word 'perception' everything man does in life is implied, including mathematics, since we can only act by perceiving the true nature of the physical world; I am using the word in a sense akin to 'experience'. The problem is pure mathematicians normally are permitted to imagine things to satisfy their nostrums, so that they do not rely solely on their percepts alone. However outrageous, they can defy reality, even gravity, logic and common sense, and leave it to the applied mathematicians, to find out whether what they have assumed is really there in nature, so that their theories based on them can be seen as true or not. In no other profession is this sort of thing allowed. Even one of the greatest mathematicians Britain has ever produced, Professor Sir Arthur Eddington, criticized that common mathematical tendency in his book, The Mathematical Theory of Relativity. I have quoted him above in the text, but it will do no harm to repeat it as it is vitally relevant here. He said: "The pure mathematician deals with ideal quantities defined as having the properties which he deliberately assigns to them. But in an experimental science we have to discover properties not to assign them..." The principle of mathematical equivalence should make them think of the practical consequences of their imaginary properties, although I doubt it, but that is another matter. The rule is that mathematicians should not seek to make the basic features of nature what they are not quantitatively, or cannot be physically; any such propositions are bound to falter. Note that we are talking only of basic phenomena. By the very nature of man, it seems everybody can make qualitative/physical changes in peripheral nature not quantitative changes in the fundamental aspects of nature, and time is the second most fundamental feature of both nature and life.

The principle means that, in effect, one cannot use mathematics to state, say, that there are ten trees in a field, and propound theories about them if, in actual fact, there are only two. This is slightly different from assigning imaginary properties to nature. It is different because it relates to 'quantities'. Six into four won't go, or something like that. The principle of mathematical equivalence rules that, to accord with physical reality, one can only talk about two trees, or as things are not as the mathematicians want them to be. Nature is not there for the convenience of mathematicians; it is neutral. That was the advantage we gained when the ancient teleological interpretations of phenomena was discredited. Therefore this rule is not to be scoffed at. I regard it as one of the strictest doctrines in logic, metaphysics and science. Science means logical thought in physical applications; metaphysics is logical thought in abstraction and mathematics is logical thought by means of symbols rather than language to facilitate the handling of size, weight, distance, volumes and complexities. So all four disciplines (including logic) are inter-relater.

It is not often realized how progressive is the study of philosophy. Quietly but surely, many entrenched myths from our primitive past are being discredited one by one by philosophers. One of them is teleological argument. With that and many other ludicrous intellectual fashions out of the way, it is unacceptable to regard any concept as 'compounded for the convenience of the mathematician', as Russell defined the Minkowski theory of space-time. Someday, we may get scholars writing about the many myths philosophers have discredited through their quiet researches to foster science and progress generally. So I regard this principle of mathematical equivalence as a strict and necessary doctrine to prevent mathematicians arrogating the power and right to alter nature quantitatively in the fundamentals of physical reality. We shall, and should, continue to alter nature qualitatively to our benefit---gardens, buildings, roads, cities, waterways, canals, railways, bridges, tunnels, all science (bar destructive devices), and all art, sports and so forth. They do not change nature but beautify it; but quantitatively, never. We cannot make one object two, or two objects one, physically. It is not possible realistically. Not in reality only in the imagination; to jump from the imagination to live conditions can be dangerous.

The origin of the rule will help the reader to understand it well when spelt out: it occurred to me when I was pondering Hermann Minkowski's claim to have made time and space into one entity as from the moment he outlined his theory, as previously quoted, in the following outrageous (even cheeky) statement: "The views of space and time which I wish to lay before you have sprung from the soil of experimental physics, and therein lies their strength. They are radical. Henceforth [that is, from the moment of his lecture] space by itself, and time by itself, are doomed to fade away into mere shadows, and only a kind of union of the two will preserve an independent reality". This is to combine two things in nature into one with mathematics ('a kind of union of the two...') It means he knew they were two independent aspects of nature. How could he have made them one from the very moment of his lecture? (Yet mathematicians continue to accept his formula as true.)

He spoke of experimental physics. In fact, the only experimental evidence pointed to time being 'local' in nature; and Einstein adopted it in his special relativity; it's the Lorentz t1. There was no suggestion that time had been found to be inextricably intertwined with space---rather the suggestion was that time could not be had without space; and that once you have space, you can create your own local time. What Einstein did was to interpret local time to mean "The only Time" we can have.

The actual physical reality known to be in existence was precisely as Minkowski himself stated it---namely, that time and space were two separate things. But it is interesting that he sought refuge in experimental physics. In that sense he did not breach the principle of mathematical equivalence. It shows that he was really a very good thinker; he had to be that good to convince Einstein to adopt his formula for general relativity, which came ten years later. The unfortunate thing for Minkowski and his followers is that the evidence he cited was really irrelevant to the claim he was making. He needed physical support that time and space are inextricably intertwined and therefore constitute one entity. The evidence that had been discovered by Lorentz and Einstein was that time was essentially local in nature, leading to the supposition that 'there are as many times as there are bodies', and that, additionally, time is different in different places, and also under different conditions. The principle of mathematical equivalence can be used to refute Minkowski's claim to have made them into one entity as from the moment of his lecture.

The rule stipulates that he could only have spoken about time and space as they actually were in physical reality, which, he admitted, were two separate entities. The reality before Minkowski was that there was space, and there was time. Even the great Einstein himself made them independent in his special theory of relativity. So it did not surprise me that Professor Sir Arthur Eddington and Bertrand Russell described the Minkowski proposal as arbitrary and fictitious. However, it did surprise me that mathematicians ignored this strong condemnation to claim that they could not understand Einstein's ideas without the Minkowski fiction.

That made me sit up and think, think of a principle to require mathematicians to relate their suppositions to exactly the nature of physical reality laid out before them, not as they would wish it to be to accord with their nostrums. I came to the conclusion that mathematics can only mirror reality, not to alter it with mathematics alone. So the principle of mathematical equivalence is this: Mathematical statements (or equations) must strictly accord with physical reality. That is the true meaning of the term "equation". It means no mathematical quantity can exceed or reduce what the actual physical quantity is. No mathematics can make one thing two, or two things one, without physical divisions and unions. Minkowski failed because, as Professor A.N. Whitehead has pointed out, time and space still pass through nature as two entities, not one. Of course, Professor Whitehead did not know what we know now, namely, that time does not actually pass by physically, but only by means of the procession of its units as created by man with his mathematics out of the orbits of the sun. Let me remind the reader once again that the theory is that the sun gives us our time units physically (otherwise we couldn't create the SI of time with units that multiply to coincide exactly with a full orbit of the sun). As these units of time proceed successively, the time is passing by. Since time's intrinsic nature remains unknown, the passage of time cannot be explained in any other manner in logic or scientific thought; but the SI works so it must have something to do with the true nature of time, surely? This is not fantasy. The real irony of the situation (which makes me smile to myself happily in self-satisfaction), is that the mathematicians tend to be religious; yet the creators of the SI of time helped to accelerate the demise of religious ideas about all time, the end of time, end of the world, and even of the universe, because by the mathematical SI of time, time became physically based, secular and traceable to scientific thought.

Yet, despite all that, the serious matter is that Minkowski rules the world from his grave, and I think that is distorting theoretical physics, even though mathematicians take mischievous pride in the situation out of spite for philosophers and logicians. The term 'space-time' is everywhere taken to mean space has been equated to time or vice versa, generally meaning space and time constitute one entity as Minkowski proposed after relativity and that the whole momentous 'creation' was achieved with mathematics---which would at once breach the Principle of

Mathematical Equivalence. In other words, mere mathematical symbols ("S=CT...") are said to have reconstituted the whole of physical reality. The reader must agree that if this is not logically accurate, then theoretical physics is being adversely affected. So I must be forgiven for repeating the whole debate again. After he was persuaded or coerced to adopt the Minkowski proposals, Einstein wrote in his seminal book RELATIVITY (Routledge edition, 2001, Part One, Sect. 17, pp 56-58): "We must replace the usual time coordinate by an imaginary magnitude $\sqrt{-1}.ct$ proportional to it..." To justify this in metaphysics or epistemology, Einstein wrote (pp56-57 of the same book), since, as I have stressed, mathematics cannot alter reality but only reflect it, even the great Einstein wrote, "...for in every event there are as many 'neighboring' events (realized or at least thinkable) as we care to choose...", and that, for all I know, is the crucial point. It amounts to stretching the transformation of Coordinates to infinity.

Is that acceptable for the determination of the nature of physical reality? Can we rely on human 'thinkability' to tell us the true nature of the world? To me the obvious answer is no. Therefore space cannot be equated to time or time to space, so the equation "S=ct..." is flawed not only in logic but also in the nature of physical reality it portrays, since time is conceptual and not physical or part of physical reality; that is why we can't find how it physically passes by. Yet it is true that space and time are so closely bound up together that we have no means of separating them

What is time?

beyond what I have been trying to sketch in this book. Thus to call them 'space-time' is correct in the secondary sense that they are virtually inseparable.

It means space and time, though independent, cannot be experienced separately for whatever metaphysical reasons, and so they come together in the perspective of man (the observer); and that man affects how human beings perceive reality, as Plato suggested. This is the ultimate of metaphysical reality and there is nothing we can do about it. Certain things in the world can never be explained. All we know is that we are alive because of it, and thank God for that. Of course, probably God had nothing to do with it at all. It may very well be the case that because of our convoluted way of getting our time in units from the determinate yearly cycle, it is man who had need of space in getting his time: we cannot get time in units without using points, and using points enforce the need for space. Without the yearly cycle man could never have invented a time system, because time is strictly linked to the environmental conditions created by the yearly orbits of the sun by the earth together with its rotations. That is why twelve midnight is not the time to go to the bush.

However to merge them in certain mathematical equations, where necessary, to be logically accurate, we must revert to the equation 'S+ct...' otherwise known as the 3+1 formula. But of course mathematicians will never do that just to please philosophers, and yet nothing will go wrong because time is always the same, what is sacrificed is the true nature of reality. There is no doubt that man affects how we see reality, because time by which we do and know everything is human in origin. To invent clever mathematics to try to demonstrate that it is not so is to distort reality.

Appendix III: Why Space on Its Own Is Not "Space-Time"

In Einstein's special theory of relativity, we learn that, "In the absence of gravity, space and time are distinct entities. In the metric of special relativity they play distinctive roles." Nothing in special relativity has changed

What is time?

since then to make all space "space-time". Yet in all their suppositions cosmologists and astronomers always refer to space as space-time.

Let me set out the facts as they are at present, as argued all through this book, and hope they will see the light. To begin from the very beginning, the whole (contentious) debate about space and time began with the work of H.A. Lorentz; until then space was space and time was time. It is true that in special relativity Einstein made space and time dynamic rather than the Newtonian absolute; but being dynamic merely means they are changeable under different conditions. But about time alone Einstein avers that he was able to complete special theory of relativity five weeks after he gained the insight that the Lorentz idea of 'local time' can be defined as 'time, pure and simple'. So let us examine the Lorentz notion of local time.

H.A. Lorentz found that time runs slower when in motion, known as "the dilation of time as a measure of moving clocks". He could not understand why and literally put it aside. He called it 'local time' or t_1. To him it was not 'the true time' but a mathematical auxiliary or curiosity--- not very important. Time, he said, was time, denoted with t, and t_1 was something you get as your local time, but certainly not applicable in the outside world as time, because it was a mere mathematical curiosity. May I remind the reader that all this has been given in detail in the text above? I have even mentioned Lorentz's own statement that he thought he failed to discover special relativity because he did not regard time dilation as important.

Strangely, however, as one of his brain waves, Einstein worked this into his theory of frames. The dilated time was 'local time'---the time of your locality. Now, if the universe was fragmented, then local time would be somebody's time, which to him would be running normally like any other time, but to outsiders would be running erratically (or slowly, in this case.)

In actual fact, that was the case with the Lorentz discovery. People outside the moving clock saw it as running slowly; but those carrying it in the moving vehicle noticed no difference in its performance. That is the genesis of the Einstein theory of frames. Otherwise time was separate from space. What you will find is that it varies under different conditions, simply because everybody has to have his own 'local time' in his locality or inertial frame. But since time is continuous, and having made it a separate co-ordinate in the study of phenomena, dynamic space would have different time co-ordinates at every turn. We recall that Bertrand Russell has stated that from the sun's point of view the tram never repeats a former journey---because the time co-ordinates would be different. Since time is a separate co-ordinate in the determination of physical reality, different time co-ordinate implies a different situation, different physical reality.

This was the state of affairs when Hermann Minkowski came in with his theory of 4-D geometry making time part and parcel of space---all space. So that cosmologists and astronomers call his theory "The Minkowski Universe", meaning that all nature is subject to the 4-D geometry, where time and space constitute one entity. But let us swiftly add that the foremost mathematical interpreter of relativity was our own Professor Sir Arthur Eddington, the man who confirmed the general theory of relativity. He wrote the definitive book on relativity, called The Mathematical Theory of Relativity. About the Minkowski 4-D Geometry, he stated clearly on Page 9 (Ch. 1.1.), as already quoted, "Such a mesh-system is of great utility and convenience in describing phenomena, and we shall continue to employ it; but we must endeavor not to lose sight of its fictitious and arbitrary nature." He was not the only great mathematician who described the Minkowski formula as arbitrary. Bertrand Russell also said it was based on arbitrary assumption. He made it plain that because of that the derivation of the Minkowski 'interval' as time from space was not valid.

Let me try and explain again the reason mathematicians still adore the Minkowski theory---even though they know that it is fictitious. It makes things easy for them. Yet it is not true. They accept the novel Einstein notion that time must be made a distinct co-ordinate in the description of phenomena. The problem is that at the same time Einstein made all time (any sort of time) 'local time'--- the time you create for your own local purposes, as Lorentz had discovered. Einstein extended the Lorentz idea to all nature. With the universe being fragmented, it was impossible that one system of 'dynamic time' (as opposed to 'absolute time'), could apply with equal validity to all fragments of the universe. As a result he said there are as many times as there are bodies in the universe. Nobody can contradict Einstein on this matter. But mathematicians found that creating your own time to add to phenomena to acquire concepts of physical reality puts too much power in the hands of mankind. (I suspect there are religious sentiments in this.) Besides, it was complicated. The Minkowski system was easier; you just have to mention the Minkowski space or ds2 and move on. It comes with time already embedded in space as part of it---so the whole of space is 'space-time' and every time is also 'space-time'. The caveat of Professor Eddington was quietly ignored. Soon everybody forgot about this; Eddington and Bertrand Russell were dead; and there was nobody clever enough to notice the discrepancy and question them about it. Of course, that leads to a distortion of relativity, but mathematicians are the arbiters of truth in mathematical physics and they were the ones benefiting from the Minkowski theory, and therefore preserved it. Otherwise it is not true that all space is 'space-time', while all time is also 'space-time'.

Yet it is true that time is always space time. You cannot have time without space; not because the space comes with time inside already, but because all time is known and used in units and units only, which can only be had by the application of points to space to create the time intervals as "relation between points". There are elements of time in the mind as the internal sense of time, known as the sense of duration. But we have got to link duration to external cycles to give us usable time in units, as I have explained above. For example, without space we cannot have the year; yet the year is our basic unit of time out of which all other units are derived. This brings a little complication but nothing serious. The reason is that you can only create time, as 'intervals', or as 'time units', as I suppose (because the year is only one unit of time and we derive all other units from the sub-divisions of the year with points or mathematics), with the application of points to space, thus making time a product of space, and therefore 'space-time'. The truth of the matter is that you cannot have time without using points to divide space; it makes time necessarily discrete, being the product of points. Therefore time is always 'space-time, or properly 'space-timed'. But that is all the connection between space and time, except that space is required, again, for displaying time in units as we have in the clock. The clock, any clock, does not give 'flowing time'. It merely reproduces units of time programmed into it. The old mechanical clock based on coiled springs gave the best illustration. The springs are manufactured to release units of time: second, second, second. If one failed to rewind the springs, the clock stopped ticking. The springs provided the clock's energy, but were strictly programmed to reproduce time in specific units only.

After the time is derived in this way, it becomes separate from both the space and the points used in creating it. That is why Einstein made them separate entities in special relativity. For, apart from the condemnation of the Minkowski 4-D geometry which assumes that time and space constitute one entity by Russell and Eddington, Professor A. N. Whitehead has also pointed out that time and space still pass through nature separately---not as one entity. To add to these, I have humbly suggested the Principle of Mathematical Equivalence above, which can also be used to denounce the Minkowski arbitrary and fictitious formula.

Appendix IV: The Misconceptions of Time in Relativity

It must not be supposed that the problem of time in relativity has been conclusively settled. Relativity is physics. When a problem is solved in physics the solution is always clear, precise in mathematics, and universally applicable; but time in relativity at present is very vague, neither definite nor precise, not least because consideration of time is a philosophical inquiry, and a very serious one too.

The arguments here are that the original Einstein theory of time can be used to solve the passage and continuity of time. Unfortunately, Herman Minkowski made the question of time in relativity immensely complex and vague, not at all like the original notion proposed by Einstein. Indeed, as a result, the question of time on the whole is destine to keep the philosophers busy for several centuries as their nostrums become footnotes to Einstein instead of Plato. As regards the physicists and cosmologists, as opposed to the philosophers, they believe that the Minkowski theory makes things easy for them; the problem is that it is just not true of the physical world.

Bertrand Russell has said the concept of space-time is perhaps the most important theory Einstein introduced. To me, there is no doubt (no 'perhaps') about it. It is the most revolutionary theory in human history simply because time is second in importance only to life itself---and yet that life cannot even be lived as a well-organized existence without time. That is how momentous time is in human affairs; and Einstein has shown that it is very different from what it has been traditionally assumed to be. Secondly, he insisted that it should be taken as a separate coordinate in the study of phenomena. In the determination of physical reality, because of Einstein time is a co-ordinate in its own right just like the height or length of matter and space are, thus making Man, the observer, part of the observed, since he has to add the time in the 3+1 formula. Those mathematicians who assume, on the Minkowski theory, that time can be incorporated into space with mere mathematics so that we can dispense with the 3+1 formula and the metaphysical role of man in the determination of physical reality, are contradicting Einstein, which is something approaching a hanging offence in science. On the contrary, it is possible that the passage and continuity of time can be conclusively resolved with the original Einstein theory of time as space-time, or local time.

There is obviously fear in some quarters that time cannot be something we invent by ourselves. Of course, if

What is time?

'there is no longer a universal time' we have to find out how we get our time. However, nobody is claiming that man invented the whole of time. Rather we have found that we invented how to quantify time by linking the natural sense of time as duration in the mind to external cycles. This sense of duration of anything is obviously connected with the memory mechanism for the retention of images and concepts in the mind.

Let me stress again, and more strongly, that the sense of time is duration in the mind. In his Mathematical Theory of Relativity, Professor Eddington made this absolutely clear, as quoted above; and we have got to take that view seriously because the theory of time outlined in this book is based on relativity. Unfortunately the mental sense of duration is not enough. It cannot give time for general use because it is private. The word 'time' is meaningless until it is objectively quantified. We need time in units to apply to the external world---i.e. to mechanize in the clock for general use, so as to be able to tell 'How much time' at a glance---see Appendix I above. This is achieved with external cycles, the most basic of which is the earth-year out of which all other units of time are derived with mathematics. And it is maintained that this is in complete conformity with the Einstein notion of time, and therefore incontrovertible. Above all, it is the only means by which we can logically solve the problems of the passage and continuity of time.

For now, we are told in all earnestness from the discussions above that relativity is not properly understood.

This may be so. But actually relativity is only a theoretical system, a suggestion. It is based on the suggestion that physical reality is not homogeneous but fragmented, and therefore subject to different natural laws. This applies to both special and general relativity. Bertrand Russell called it 'a logically deductive system'. In plain language, 'a new philosophy of physical reality' so logically structured that it demands attention, respect and serious study. And these Einstein has certainly achieved. With Einstein alone we are not talking about genius but a godlike intellectual phenomenon never seen on this planet before; he reconstructed the world of physical reality single-handed, that is the reason he is indispensable to both scientists and philosophers.

So Bertrand Russell was absolutely right. Einstein's system is a new logic of physical reality, and it works. But theoretical physics is most unlike the physics we apply in laboratories. Ordinary physics is much more like chemistry; it has consequences. The Nobel Committee was right to award Lord Rutherford the Prize for Chemistry, even though he regarded himself as a physicist, who had rather cheekily claimed that "all of science is either physics or stamp collecting"!

In theoretical physics there are no obvious consequences, so it is difficult to judge the merits of suggestions. Instead, when we get a new theory in advanced physics (rightly or wrongly), three things will happen. I mean, all three will definitely happen in succession, whatever may be the merits of the new

What is time?

proposal. First, we will get interpretations of the basic theory proposed in such complex settings (or confused formulas) from rival theorists that the debate just has to go on; nothing will be settled in the meantime. But because there are no consequences, nobody will get hurt, no machinery will fail to function; avoidable calamities will not occur. The rains will not stop; the sun will not dim.

The most recent example was the eather debacle (or debate). Secondly, we will get accusations and counter accusations of misrepresentations and misunderstandings. The third possibility (because philosophers share with theoretical physic one subject-matter, being the determination of physical reality), will be philosophical interpretations to arrogate the almighty right to shame and discredit some of the factions in the debate, only for philosophers of different schools to turn the tables---and so the debate will be carried on and on. These philosophical discourses are often pretty profound, giving several intelligent interpretations without being able to settle the argument one way or the other. Strangely, that is how we eventually acquire our knowledge of the external world, sometimes referred to as the practice of 'academic freedom'. That is what happened to Plato. And that is what is happening to Einstein as he has come to replace Plato, in fact, to make his basic suggestion redundant, if not completely false, due to the quantum theory.

A careful examination of what has happened to Einstein's theory of time so far betrays elements of all three conditions. First, we are told that 'most definitely' due to

Einstein's analysis of 'Order and Simultaneity' there simply is no 'standard or absolute time frame in the universe'. ('Time Frame' or 'Time Reference' means the same thing. It means the logical criterion of validity.) This is generally accepted as true; for it is reinforced by the Lorentz time dilation and local time concepts.

However, it implies that time in the abstract is utterly indefinable, as I have shown above with discussions about the earth-year. The year is indefinable; other time units in use on earth are defined in reference to the year. But the year on its own is logically indefinable. Again, all time units, down even to the caesium units, are based on the earth-year; they are meaningful only as related to the year; but like the years, on their own (that is in the abstract), none of them can be logically defined. How long, for instance, is a second in logic without reference to something else? The result is that we all have to use the clock, or clocks, based on the earth-year. By this theory of time (as quantified time), the human intellect is built upon the concept of "points and instants". Instants do not exist independently in nature. Only points do; they had to be discovered by man, but they do exist in nature independently—for example, trees constitute points. Before we learned to put points on paper, we could see that trees dotted the landscape. Thus points constitute the basic instrument of human thought, especially in mathematics from which all the sciences spring. The instants arise from the act of 'consciously' and 'purposely' moving from point to point, confirming the Russellian notion that time is 'relation between points'. Hence quantified time is human in origin, except that the internal sense of time (as duration of anything in the mind) must be recognized as making a psychological contribution to the invention of quantified time in that the external cycles used for quantified time (the years, for instance), have to have psychological anchors (meanings) which are the sense of duration of anything in the mind.

Secondly, in the absence of a standard time frame, what does it mean to claim that time intervals in a moving frame are shorter---shorter as against what kind of standard or universal time? What time intervals are they compared with since there is no standard time frame? (Note that you cannot say they are shorter as compared to other clocks outside the moving frame; that will bring in the Einstein theory of frames, as I will discuss presently.)

So we all, in the end, have to resort to using the clock or clocks based on the earth-year. Yet if we use the clocks then it is not correct to claim that time intervals in a moving frame are shorter; they are not naturally or normally (in its proper setting) shorter or longer; they are normal to that frame, or to its natural frame. The moving clock may only seem 'different' as viewed from the outside; but if that is the case then there is no puzzle. The time of the moving frame is not 'our' time; and it is not queer to its natural environment or setting. It is a strange phenomenon to those looking in from the outside, in breach of the Einstein theory of frames. In fact, it is irrelevant to anybody but those in the moving vehicle only.

The whole idea of studying other frames from the outside is fraught with difficulties; it can never be an exact science since the standard postulates that make our system work (and make it what it is) might be inapplicable outside our frame, or planet. Speculations into other frames from our frame have been responsible for all the bizarre suppositions about time and space-time from mathematicians and cosmologists in general relativity. I don't think that kind of enterprise is justifiable, especially when it leads to theories that space-time may be infinite in its timelike directions. Space-time cannot be infinite because it is necessarily discrete---the year, for instance, is not infinite. It is only one; all other units of time derived from the year are also discrete and individual. The proper way to think of time as space-time is that its units are in perpetual procession (one year or second following another) to make time seem continuous; as such time can never be infinite.

Nothing illustrates the confusion about time in physics as a result of relativity and how it is misunderstood by scientists than the story of muons. By normal logic they should not last long enough to reach the earth; but they do. With the use of formulaic mathematics and concepts, physicists explain this by saying special relativity provides the answer as follows: the speed of muons is so great that their internal clocks slow down. Using the theories of time dilation and the so-called twin paradox based on it, it is assumed that as the muons gathered speed and their internal clocks slowed down they aged less and thus are able to last long enough to reach the earth. To a logician or philosopher who understands relativity, this is so laughable as to choke him. It is really the best example of the confusion in physics about time in relativity. (1) Time dilation has nothing to do with the muons and how they behave, since time does not dilate internally. Lorentz found that a moving clock would be seen by outsiders as running slowly; but internally those carrying the moving clock would notice absolutely no difference in its performance. Einstein explained this with his theory of frames---the moving clock is in a different frame. There is no logical mechanism for this kind of episode to be able to control time per se. All other clocks would not run slower or faster; and since there is no such thing as an absolute time frame, or standard time, by which all other clocks can be compared, the moving clock's performance has no relevance at all in physics, because its carriers would notice no anomaly; and those outside who notice any anomaly should mind their own business since it is not their time. (2) The idea that muons have internal clocks is based on the Minkowski theory of space-time, where space and time are assumed to constitute one entity; and therefore the reasoning goes that, since the muons occupy space, and all space is space-time, they have their own internal clocks to keep or measure time for them. Again, any logician will describe this as nonsense; for after all, the Minkowski space is known to be fictitious and arbitrary with absolutely no logical validity.

The basic idea in Time Dilation, which these writers rely on, is easily disproved thus: we know there are (roughly accurately) specific times by our normal clocks for the occurrences of certain events on this planet. Let us use Sunrise and Sunset for illustration. If Sunrise is usually 6 am, and Sunset is roughly 6 pm, as they are in some countries in the Tropics, it is inconceivable that a moving clock can force or influence these times to become 7 am, and 7 pm, on the planet all over just because one particular clock somewhere is running an hour late. "The dilation of time as a measure of moving clocks" can in no way influence all time per se on the planet. It affects the performance of only one clock. Clocks are manufactured to reproduce specific time units, usually in seconds. If a particular clock, for whatever reason, is running erratically, there is no logical mechanism for its behavior to affect all other clocks on the planet.

The reader will have noticed that the name of Lord Bertrand Russell comes up regularly in all discussions of relativity's interpretation. It is inevitable. Russell was highly respected by Einstein, and for very good reasons. He was the world's greatest philosopher at the time. He was also a great mathematician and logician of genius. A most attractive writer, who won the Nobel Prize for Literature, he wrote about every subject in philosophy, including novels to illustrate moral points. When relativity was announced, he abandoned many of his most cherished ideas as wrong without shame or even mild embarrassment. He was candid and honest in the most adorable way, completely dedicated to the truth no matter how it reflected on his own beliefs. Russell probably had no certain beliefs other than the pursuit of the truth wherever it took him: via science, logic, mathematics or plain common sense, and linguistics. If he was certain that teaching mathematics to people from the cradle could save the world, he would have advocated that as his philosophy.

Concerning relativity specifically, in the later editions of his little book "Problems of Philosophy" he denounced his original philosophy as expressed in the book because of Einstein's theories, joking that whoever wrote the original ideas must have been a monkey, but nobody should suppose that the monkey looked, even remotely, like himself! No great philosopher has ever made such a confession; often associated with rulers, they all wrote imperious edicts as if they had discovered the final truth in logic and metaphysics. Indeed, Russell later called his Fellowship dissertation "somewhat foolish" for the same reason, namely, the geometry used by Einstein had made his discussions of the foundations of geometry completely wrong, and he was happy to admit it and adopt the new Einstein theory. He wrote one of the best interpretations of relativity, still in use, under the title "ABC of Relativity". His book "The Analysis of Matter" can be divided into two. One section is about relativity; the other is mainly about his joint theory with A. N Whitehead to the effect that the world of sense is a construction, not an inference. Yet even this can be traced to relativity, since Einstein made man the observer part of the observed, meaning that man contributes something to the nature of physical reality---i.e.

to help with the construction of that reality---and the book was published long after both special and general relativity. It is a moot point.

But can we suppose that Bertrand Russell put an end to philosophy as an academic subject except in the form of 'The philosophy of science'? Let us try a little logical analysis here. Time has no existence outside the human mind, which, according to Russell, constructs it out of the elements of nature; yet without time nothing can come to be---it needs time. Time underlies all reality, all existence and any concept of creation. None of these can occur without the human mind's creation of time by which alone we can deduce existence or reality through scientific principles. So when we're all gone, swallowed up by a black hole, apart from chemistry and the perpetual flux in nature, there could not be reality---therefore man appears to be the creator of reality. This calls for humility, kindred love and mutual assistance as the best religious ideas. So, even though I am an unbeliever, I think we do need religious myths in this hollow and lonely life---at least for those who cannot live without them.

Conclusion

The repeated definition of time in this book is that it is conceptualized in the mind as a period of waiting from external contacts of any kind and number of images, and does not come to us physically from anything cosmic or otherwise. But every book about time is a thesis about life too because time and life are bound-up together and seem completely inseparable. Being such a serious subject, I feel obliged to spell out the scientific, logical and philosophic grounds upon which my ideas are based; I further believe that they must be postulated at the beginning of the conclusion, even though they have already been stated or implied in the book more or less clearly. Thus I list below the five principles I have relied upon, which I know to be generally acceptable. Of course the Conclusion is long, for time is a difficult subject---I offer no apologies.

1. As has become the practice in our new world of the quantum (or QED that we owe to Einstein), the revolution

What is time?

began with Albert Einstein's adaption of the Lorentz discovery of local time, by calling it 'time, pure and simple', even though Lorentz himself had no idea what he had discovered---namely, that time is variable and not at all fixed so that a second here is a second everywhere else. Yet that was the moment the whole universe (from our human perspectives) changed from one of myths that we can imagine to one of systematically searching for the truth consistent with everything we know as 'Created' by man (wherever man came from), because time controls the whole of human existence and everything we do, yet it became scientific from that unique Lorentz/Einstein moment of cosmic genius.

2. The next step was rather simpler but the question it posed has never been answered. People just use time and forget about its provenance, other than claiming that it just is there. But if we agree, as we have to due to the experimental evidence, that time is not fixed, general and cosmic (even without mentioning the toxic word God), then what is it that the clock measures for us as time? Bertrand Russell was the first to ask the question, and I think I have provided a credible answer in the book.

3. To answer the Russell question it was necessary to investigate how we get our time, the result of which is "The Logic of Time in the Universe", applicable to any 'Beings' anywhere in the universe since all reasoning is based on the humble syllogism, so that the logic of time becomes universal. The logic, not the time. Time may not be

universal, but the logic of how to invent time is bound to be universal.

4. This kind of time is also bound to be discrete because logically it can only be based on repetitive cycles or motions, and such determinate motions can only spawn discrete units of time, intervals, or anything else that relies on points: it will have 'a beginning and end' scenario and no more. The best example is how we get the years.

5. Discrete time cannot move; so time can only advance through replications: the units of time in perpetual procession cause the illusion of the passage of time through the universe, even though we know that universal time does not exist ---e.g. the years increase in numbers to pass by; and that is how the passage of the years is achieved to give us our age-numbers, since time is only conceptual and so cannot be capable of physically passing by. It is because for centuries time has been regarded as a physical entity that its passage appeared mysterious. As a concept it can only pass by in the mind through arithmetical calculations. Any good thinker analyzing these principles will, I hope, arrive at just about the same conclusions as argued above and detailed below.

Time is so familiar that everybody thinks he or she knows what it is; in any case even those who desire to study it come with their own agenda of what it is or should be. Most of this knowledge of time has come from tradition and custom. To even try to sketch them will require a tome; there are an infinite number of ideas about time. Albert Einstein showed that even time in the clock is not what we

What is time?

take it to be. We owe this new theory of time not entirely to Einstein. The original idea or discovery that time is changeable came from the researches of the Dutch physicist Anton Lorentz. Previous to the new theory all mankind accepted what is generally referred to as "The Newtonian Absolute time." In fact, it wasn't Newton's idea alone; it was really a tradition whose origins are buried in religious tombs. The essential features of this time was that it's universal, generally covering the whole of the universe (because, of course), it's supposed to be divine in origin— and God was the "Creator", you may recall. As such it's also fixed; no human being dared to suggest that he or she could alter God's creation. Being fixed meant it was the same everywhere; that every unit of time here is the same everywhere. Logically, every schoolboy will now dispute this idea, for at least we know that the earth's orbit of the sun is what we call one year, our basic unit of time; but there're several bodies competing with our orbits, and we can guess that they take different periods to circle the sun, How can one unit of time based on the year be the same everywhere? Such questions influenced Einstein to make him declare that Lorentz's discovery of a different sort of time (from the traditional one) was rather a discovery of the real nature of time. This was not only amazing; it's revolutionary---even shocking---and freed mankind from the oppressive constraints of absolute time. Looking back, it's astonishing how slow knowledge of the external world tends to spread. The orbit of the sun is used for time and yet we could not connect that with the units of time we used and believed that they're independently created by

God, were fixed and generally permeated the universe---how presumptuous. But then we have to remember that at one time even our tiny earth was supposed to be the centre of the universe! Man may be small, but he has brains that seek to soar above the stars. And the clerics were the happiest for they exercised all this awesome power over the universe---how very presumptuous and vain, no wonder they invented the afterlife to come back after death, because life was good!

Since Einstein we have come to realize that everything we normally refer to as time (meaning every unit of time), is derived from all the planning we are able to do with the 24-hour periods of the revolutions of the earth, or the bigger 12-month round trip of the sun. Examples are everywhere: time to catch a bus; time to go to the shop; time to go to school; time in sports; time for doing all the things we do. Another peculiarity is that we know time only in units. Some people believe that there is such a thing as 'silent time'. In sleep they claim that time is going silently. Well the clock is going anyway; but without the clock, is time going still? The positive answer is that the earth is moving us to different and new positions in the cosmos. We can extend this idea to chemistry. Motion and chemistry can cause us 'a period of waiting'. That's time, of course, but how do we tell 'how much time it is'? I have explained this in the section entitled Time and Quantified Time. We quantify time into the specific units familiar to all of us. I argue that without quantification the word time has no precise meaning because it can be caused by anything---

What is time?

motion, chemistry, inertial, gravity, entropy, etc. These can all provide a period of waiting as time. What is culturally useful as time is to be found in the clock, but how did it get there? It is based on quantified time, any cycle at all counted as units of passing time.

The problem is that when we quantify time it becomes discrete. Yet all time has to be quantified (that is, reduced to units) to be useful in society, otherwise mentioning the word 'time' has no meaning except a private one to oneself. The old religious idea was that time existed all through the universe and we used our mathematics to create suitable time units for ourselves out of the blanket, universal entity. Once that notion was discredited, we had to find out how we get our time in units. But the process of quantification renders it discrete. Discrete time cannot run through nature, yet everybody refers to time as if it is running through nature. The old traditional view of time seems to me completely impossible to expunged from the human mind; we are wedded to a notion of existence linked with time as something like a thread running through nature and the mind can't free us of that. My opinion is that it is because once time is known it becomes part of existence, since one cannot define existence without 'when' it's there; our nature is bound-up with the old definition; the new explanation may be more logical, but the old one is proving hard to expunge.

If we accept that we know time only in units, specific periods or moments, it implies that we also accept it as secular; for since the parameters for creating any inertial body's time units are different, every inertial body has got to invent its own time system. This is what has become known as 'Secular Time' in place of the old idea as "a fixed time system of divine origins covering the whole universe." Secular time can be logically traced from the parameters in our environment, meaning we create it here and will be applicable to us alone. As stated by Einstein (quoted above): "There are as many times as there are inertial bodies". And the greatest philosopher of the time, our own Bertrand Russell, explained that our time is actually constructed as relation between points. For it should always be borne in mind that no one person, however clever, can decipher all the intricacies of time built up over so many centuries from before we're human to when we came down from the trees. So even Einstein required the help of the greatest philosopher alive.

Unfortunately, time is so closely associated with life that they don't seem to be separable; as a result we're happy (the mind is clinging to the old notion because it is 'happy') interpreting it as divine. Yet that's because life was also supposed to be divine. That God created life and the time to go with it. When that religious idea was rejected (or had to be abandoned in the face of scientific evidence: Darwin, astrophysics, scientific medicine, electricity, etc.), the Russellian query came in, showing how important philosophers are, namely, if cosmic time is abandoned, what really is measured by the clock? Part of my frustration is that nobody is interested in answering this question but

What is time?

rather wish to go on regarding time as if it is still running all through the cosmos and the same everywhere.

However, frankly, not everybody accepts the secular theory of time. The religions have been completely defeated and they know it. They are pretending to be staunchly sticking to the divine origins of time because all of its mysteries cannot be laid bare by science; and they get away with this deceit because the whole of mankind is still behaving as if the new secular view of time has never been thought of at all. And they are able to do so mainly because time is reckoned in the same old manner we've become accustomed to over centuries, that is using the year as a unit of time and paring it down to the seconds. It is the basis of secular time; but it has been the foundation of religious time as well over several centuries. Is there a mystery here? I do not think so. We've always used the earth-year for time, hence our notions of 'years' to account for ageing. But the view was that we're merely calculating the year with our mathematics to accord with divine time running through the universe and of which the year is part. It was not accepted as an original creation (construction as Russell put it) of time without divine intervention, which is what the new theory of secular time amounts to.

And we know that the new secular theory of time has to be true otherwise science wouldn't work; for this new assumption is proved by the number of quandaries that have become easily explicable on the secular theory of time derived as a construction based on "relation between points". Some of these issues that the secular theory allows us to formulate in consonance with science have been discussed in the book. They include the following: (1) That the past, present and future syndrome is memory of past events plus the consequences of these events carried with us to the present and whose remainder or results would be carried with us to the future---that history is seen as the march of events not time. People cause events, not time; the times are associated with the events to show the times and dates of occurrence; and obviously today's events have antecedents (the past), as they will have consequences (the future or as the future). Einstein was therefore absolutely right in calling the syndrome an illusion. Of course it is true that time can cause events but only temporarily, or accidentally (like an avalanche going off after a while under the weight of additional snow), not as a continuous story. The continuous events of history (telling a human story), cannot be caused by time as a period of waiting. History is caused by sustained action deriving mostly from the past and often going on to cause the events of the future too. Time might even be irrelevant. For instance, soldiers in the trenches fighting over several days might not even notice that the old year had given way to a new one, and that Christmas had already gone.

(2) It is also easy to explain time travel on the secular theory of time; for the essential feature of that time is that it is discrete, not in a chain or a tread but proceeds unit by unit---exactly like the year. Time cannot be reckoned in any other way (this is one of my assertions and I'd stick my neck on it), for the real time is never known. What we call time is

What is time?

the number of units we create with any cyclical motion at all. Over the years, we've realized that the earth-year provides the most convenient cycle for reckoning time---because of the astronomical features. But that should not mislead anybody about the nature of time. Orbiting the sun and calling each orbit a year and sub-dividing the year down to the seconds may be more convenient, but it's not much different from tapping the finger and counting them as the rate of passage of time. Time is the repetitive cycles or motions we count as the rate of the passage of time. Real time in nature is either non-existent, in the absence of relevant parameters, or unknowable---unknowable not in the religious sense but in the sense that it's too complicated to decipher, involving several of the agents that can cause us to gain the sense of a period of waiting, which is what we call time.

So long as the year is seen as one unit of time that has to be repeated to continue, the time system based on the yearly cycle cannot be otherwise than discrete, proceeding unit by unit. In procession we get the continuity of time. I was rather surprised this was disputed by the very people who usually celebrate the end of the year so enthusiastically, so much that they deliberately engineered the birth of Jesus to occur at the very end of the year---crafty saints! You have to ponder it carefully to realize that it was a ruse; otherwise, as they wanted, you'd think it was real. Meanwhile the Christian soldiers were marching on, the colonial exploiters were moving in, and the instruments of oppressions were being assembled. What is wrong in the world today started long ago in history and have gained such roots that the world will never be a peaceful place. I'm afraid we've lost all hope of a rational, well-organized, peaceful existence.

(3) Time travel has become popular in recent times; publishers are outbidding each other to publish books on time travel, because the man reputed to be the greatest logician ever to grace this planet, Kurt Gödel, has claimed that his conversations with Einstein has convinced him that time travel is 'a scientific possibility'. On the other hand, nobody has ever asked to read any manuscript of mine. The majority of publishers and institutions never even reply to my submissions. But I am not bitter for time is a very serious subject and extremely difficult to write about. The human mind loves familiarity and often attacks new ideas fiercely. A theorist is lucky if he is not thrown to the lions as the initial reaction of the powerful to the unfamiliar; that they often turn out to have been correct but ahead of their times is immaterial. Human beings cannot act against the basic tendency ingrained in the brain. We progress but only slowly, any attempt to jump the queue could be dangerous. Better do the work and die off; the world will recognize it when the time comes, provided the work is good. If not, well, you tried. Ideas may incite strives and revolution, but not the theorist. He wouldn't even know the likely effects of what he writes. It is a defect in intellectuals and a welcome one: proposals should be advanced on logic and morality and left at that. What happens next must on no

What is time?

account be predictable or predicted. It's more likely to go awfully wrong if predicted.

As I berate the greedy but shallow publishers of books on time travel, I also realize that they might be hedging their bets; and I honestly can't see anything wrong with that. For there is one mathematical theory that seems to make time travel possible. It is, as mentioned in the book, the 'great' Minkowski formula for equating space to time. He's not great; his theory too was not important either. But mathematicians seem not to care. They want time travel and the Minkowski proposal appears to make it feasible. He said he had achieved the union of space with time so that they've become one entity. I agree that if that is true then, given the curvature of space, time could go forward or backward.

The reader does not have to take my word for it. Professor Arthur Eddington and Bertrand Russell have called the Minkowski theory arbitrary and fictitious. I'll settle for that. Since the Minkowski formula is based on imaginary time coordinates, it certainly cannot make space take time with it when it curves. I condemn the theory as false, but he came close, very close. Unfortunately in science and logical thought coming close is not good enough---our lives depend on these ideas and we might lose them if they're not really true of the external world. Those people, invincibly arrogant people, who disparage philosophy should realize that science (electricity, running water, weather predictions, geostationary satellites, building and construction, agriculture, etc.) and scientific medicine (antibiotics. Immunology, hematology, all those shinning machines and scanners in our hospitals, blood pressure and its solutions, drugs and all the rest of it), have come from the numerous attempts to understand the world and nature generally for the benefit of mankind---the point is, they all arose as a result of arm-chair speculators imagining things sometimes to the ridicule of their fellows. Science is only the material aspect of speculation; the really serious work is done by the philosophers thinking logically about everything privately. One unbroken rule is that all inquiries have to follow a logical route. The reason even the most abstract mathematical system has to have a logical premise.

We accept that we do not know the truth about nature or the external world. I have mentioned the Platonic simile-of-the-cave, the logician Kurt Gödel's Incompleteness Theorem, the impossibility of knowing what time is except how time units are passing by, and the fact that we can never answer all the mysteries that surround time. We can never know everything. But all suggestions have to pass the test of logic because our lives depend on them and we do not want to walk into bottomless pits with our eyes wide open.

Over the centuries we've evolved various methods for establishing the reliable ideas by which we live. At the top of such methods is logic, the chief method for knowing what is reliable, safe or profitable. By such rules it is obligatory to reject the Minkowski equation of space to time; and once that is rejected time travel is abolished. If his equation of space to time is not logically tenable, then time cannot travel (move, whirl, curve, gravitate in a black hole) together with space, moreover in a time system that is strictly discrete, proceeding unit by unit: it should always be remembered that time is known and used in units; these cannot be plucked from the thin air but deduced logically from a physical premise.

(4) The passage of time has troubled thinkers down the centuries. The latest proposals centre on the arrow of time theory. It is assumed that time moves in one direction only; and that is regarded as a kind of arrow pointing to the direction of time. A thousand other suppositions tell us that this arrow can be reversed to allow time travel backwards; or it can have several dimensions; it can even turn worm-like and so forth. A number of events in science are called 'time reversal' simply because direction or motion is reversed. The most questionable is what they call 'nuclear reactions going backwards'. If that is time reversal then why is consciousness stable and we are not pulled forward and backward constantly by these invisible sub-atomic particles? What neutralizes the reversals to leave existence always stable? The idea is flawed because it equates time with ordinary motion. We don't know the true nature of time; but we shouldn't interpret it in a manner we can use to prove any ideas we suppose----the religions do that and we castigate them severely for it. I think scientists are confused by the reversal of action through films or magnetic tapes. Yet action on films are just actions on films, not time for all the billions of people and

the billions of their activities on the planet. Time is when any action is routed through astronomy, mathematics, physical reality and social practice. This is a long way from what happens to any individual; otherwise we would be dominated by endless seesaws. Of course we use the earth's motions to reckon time, but ordinary motions on the surface of the planet, numbering billions, cannot constitute collateral time systems otherwise we could not have a stable time to live by. They are merely motions of uncountable objects. Nuclear reactions (or whatever else) going backwards are just 'actions going backwards' (just like walking backwards). They have nothing to do with time. Even atomic time, as I have noted above, is not different from earth time, only more accurate in measuring the second (and the second had to be a time unit, a fraction of the year), to qualify as 'time', otherwise the atomic oscillations as time would not make sense. All motions as relation between points cause 'various periods of waiting' which we know as 'time'; but there are billions and therefore unusable as time in the clock. We have chosen the earth year for reckoning time in the clock, and that is not naturally reversible. The multiple dimensions and numerous times up and down in science are philosophically laughable. Ever since Einstein shocked the world, every scientist wants to be called a genius. They forget how rare geniuses are.

In fact, time does not move at all; all the movements in the story of history are physical and involve human motions and activities; they are what tell a story, and all history is a

What is time?

story of how human beings have behaved or reacted to the environment. The year does not move from one year to two years. It is replicated to be years. So the arrow of time is not discussed at all in the book; my argument, all through, is that the new secular time is basically discrete---simply because the basic unit of time is only one; it does not stretch and it does not move; but it does expire completely. We start a new year at the end of the current one. This year replicates incessantly, hence the centuries. Discrete time passes by through the procession of its units, and that exactly is how the year passes by together with its fractions, from the second to the hours, weeks and months. The pulses of atomic time are included because they are based on the second. The radiation pulses of caesium 137 are used to measure the length of the second more precisely; that is all atomic time means; so you can't have atomic time without the year. The year remains the basic unit of time out of which all other units are derived with points or mathematics---always in association with the astronomical features of the earth and solar system. And the year is determinate; therefore the time system it affords cannot be anything other than determinate or discrete time. Discrete time cannot march; it cannot spread; it cannot curve and it cannot be the same thing as space because it is created with points as applied to space, that is the reason it is discrete---from one point to another. Thus it is 'a product' of space, part of it but not the same thing. Time can still be called 'space-time' since we cannot have any unit of time without space. But space-time in the sense that space and time are unified into one entity as Minkowski proposed is obviously untenable in logic.

Thus, the concept of curved space-time by which time travel is said to be 'a scientific possibility' has been dismissed in the book as a bogey. Mathematicians hope it were true; it would make many things in the study of time simple. But it is based on imaginary time coordinates and therefore cannot be true in nature---only in mathematics.

I have also discussed other relevant topics such as Time Dilation, Gravity, Entropy and time, The clocks paradox and Twins paradox. From the point of view of secular time, these and many other mysteries can either be reasonably resolved or regarded as insoluble mysteries. We cannot explain everything; the life itself is not understood; we're orphans because we do not know whence we came, and should live carefully or cautiously without making demands on nature, since we're completely alien to it as it is also alien to us.

My final word is that nobody can ever discover the true nature of time, even if it exists. However, we have constructed a time system, according to Bertrand Russell. And it seems to me to raise two problems in philosophy: (1) Does time move or only the cycles we use for its reckoning do move deceptively as if it is time itself moving on? (2) Is that movement what we call "the passage of time", so that all man can ever know is the passage of time and therefore needs no theories to account for how time passes by? In either case the problem of the passage of time is solved.

What is time?

The passage of time is significant only culturally; and although based on physical parameters, it is inferred only conceptually, otherwise a year, the days, months and weeks mean nothing to anybody. These arguments lead to a number of conclusions and suppositions such as these (with all the fallibility of an ordinary human being, of course):-

1. Mentally time is a period of waiting and therefore purely conceptual;

2. However it is based on physical parameters that render life safe in an inertial body. To identify such bodies accurately, Einstein's two postulates for special relativity must include time, namely that the parameters necessary for the creation or construction of time must be present, since time is vital to life and yet is not a cosmic phenomenon---or naturally there;

3. This is to acknowledge that Russell was right to describe time as 'a construction' under relativity;

4. Russell was obliged to investigate the origins of time because under relativity time is not seen as cosmic. That being so it means there is no time in the universe until somebody invents his or her time. All 'Beings' in the universe will face the same problem. I call this the 'Logic of Time in the Universe'.

5. Metaphysically time is defined by Professor A.N. Whitehead as "a sequence of non-interacting moments", or any contact with nature, I should add. For there is only one medium for knowing a moment in nature; it's through

physical contact or perception in a general sense, or any contact with anything whatsoever. This was the beginning of gaining our knowledge of the external world that eventually led to the creation of civilization. To know how long is or was any contact, we count cycles or repetitive motions---the years for instance. Tapping the figure amounts to the same thing. Thus arose the need for (and the sense of) time in sentient beings, or thinking man. It's a long intellectual journey that ended in the creation of the clock. It may even be the basis of the creation of civilization. What I feel is that it is certainly the foundation of our intellectual aspirations. Otherwise how did man establish contact with the external world or reality---the reality to provide the facts and ideas for thought?

6. However, it implies that time does not move. Contact is just contact, perception. What are in motion (and which we count to know 'how long') and also mislead us into thinking that time is in perpetual motion, are the repetitive cycles we use to reckon time---or what we count to tell us 'how long' was the duration of any contact or moment---i.e. in sophistication it is the year pared down to the seconds, or even the atomic oscillations.

7. The essential question is, if so then what is history about? The answer is that history is the story of what happens to a sentient being, not how many times the earth has circled the sun. History is the march of events not time. History is always in the past. History is a story. But time, on the other hand, is always the future of the present: yesterday is history, tomorrow is the future, and today is

What is time?

the current reading of time, or is 'time' for short. You need to know the time for a task; you won't need the time when the task is finished. As mentioned before, an intimidating team of hungry academics have produced a tome published by CUP called, "Time's Arrows Today..." The thesis of this book is sadly mistaken. Time has no recognizable arrows; it is not physically passing. It replicates its moments to continue 'successively'. There can be no reverse of time, logically, because the units are lost when replaced. Another nail in the coffin of time travel. People should by now begin to realize that Einstein was a unique genius because of his abolition of absolute time alone---yet there are a few more successes. Our own Bertrand Russell's question about what is measured by the clock was also very clever. The clock confirms that time is necessarily discrete.

8. A time system based on repetitive motions is bound to be discrete. Discrete time consists of non-interacting units or moments; as such they do not march. The years increase in numbers successively to pass by. I repeat, the year out of which all units of time are derived as fractions thereof, does not march on and on as one single year notching up miles in space. It ends on 31st December and another begins. The problem of the passage of time never existed. Properly defined, time does not pass; it also has no direction. Rather its discrete units are in succession to make time seem to be passing by. And the discrete nature of time is not in dispute, because time is based on the yearly cycle and that is determinate, occurring one after the other successively.

9. Actually what we know of time is only the means by which it is passing by---i.e. through the motions of repetitive cycles, examples being the passage of the years, the minutes, hours, et al. The passage of time is the time, the succession of the units, since no time unit stands still. The truth of time has always been evident within the mechanics of its construction; except that nobody could think of it, given the constraints of the religious absolute time foisted on man. Once we realize that it is constructed, the logical deductions of its true nature becomes clearer. The year is our basic unit of time from which all other units are obtained as fractions, but all we know of the year is how it is passing by.

10. In my opinion, Leibniz was right to suppose that time is a succession, and to his credit he said this more than three hundred years ago---long before Einstein showed the true nature of time. What was missing in his day was the concept that time consists of non-interacting units---thus a succession of units or moments obviates the need for time to be marching on;

11. Time can be merged with space but, logically, only through the 3+1 formula, not by mathematics alone as Minkowski proposed. There are physical, physiological, psycho-logical, mathematical and astronomical aspects of time; no one can incorporate all of them in one mathematical equation. Even in literature only philosophers can incorporate all of them in a few, credible sentences. The four-dimensional space may exist, but not including time, because time is derived from space with

What is time?

points in the first place as 'relation between points'. That is how we get the year as a unit of time which we pare down to the seconds and the totality of our time units, since even the atomic units are related to the second. As an entity produced by space with independent existence, it cannot be swallowed back by space in ordinary circumstances, but can exist in partnership or association with it. Thus all those complicated mathematics ending in the equation "S=CT..." are logically flawed. You cannot command space to become time with mathematics. Einstein was wise enough to make space and time separate in special relativity. In any case, even in general relativity we are told on the authority of Professor Yourgrau that Einstein never even bothered to understand the theory of four-dimensional geometry, and he added, "Yet, in spite of that, Einstein did the work and not the mathematicians"! I know, of course, that scientists have got themselves into a royal-type, big and kinky scandal, because if 's=ct...' is not true then a big chunk of cosmology is bunk. Time alone of all subjects does not have to be created. As a period of waiting, time will always be there no matter what the cause is, the reason all statements of time are meaningful. Problems arise when quantifications come in and the parameters used happen to be different from those found on earth. However writers on time face the frustration that time is always there as a period of waiting. That is what Minkowski exploited, but it cannot make time the same as space, and that is what Einstein cleverly realized by making space and time separate phenomena. It also cannot solve the problem of the passage of time. S=ct even if true, cannot solve the problem of the passage of time. To solve that we need to show that space and time are separate as it (at the same time) indicates its provenance. If time is not cosmic as experiments have indicated, then we are right to want to know what it is, how we get it, where it comes from and how it passes by as we think it is always doing. Intellectually, apart from the nature and origins of life, no problems are greater than these about time.

12. In the absence of 4-D geometry, curved-space cannot take time with it when it curves. 4-D geometry was supposed to link time to space and make them into one entity so that they'd move together. But it could not achieve that miracle because the basic equation relies on imaginary time coordinates. I am very surprised (not to say ashamed), that mathematicians continue to recite the Minkowski myth. Nowadays we are not all of us as ignorant of mathematics as we used to be. An imaginary entity cannot descend from the heavens with divine powers to link two physical entities into one. In this situation, space remains separate from time as we have always known it to be. The time is constructed by man; that is our best logical definition; but the space for all living things, plants and animals, mountains and rivers, is natural. Hence time travel through the tunnels of 'curved-space-time' (backwards and forward), which some science writers assume to be 'a scientific possibility', (as stated in Professor Yourgrau's book A World Without Time) cannot be anything other than pure fantasy. All problems can be solved through fantasy; what philosophers do are consistent logical deductions in

What is time?

which fantasy plays no part, so that we will know the successful results as dependable, reliable, eternal and certain knowledge. A Scientist and a philosopher do the same things, except that one has mechanical aids, but the other has only his brains.

13. Again, let me repeat that once time became known as 'a construction'---thanks to Bertrand Russell---logicians had no difficulty discovering its mechanics physically, so that terms like time travel, cosmic time, divine time, Day of judgment, the beginning of time, time warp, counting successive days on the way to the end of time, the passage of time and so forth, all became redundant. It is true that writing about time has therefore become much more difficult, but certainly devoid of fantasy, religion and cant. It has also altered the intellectual landscape permanently but that has not sunk in yet since relativity is not properly understood. Bertrand Russell too is regarded as only a rebel. For about fifty years my proposals about these issues are not even read before rejection, most of them are never even acknowledged; if they're read, surely, somebody would have noticed that the theory of time has really changed.

14. Yet because of Einstein, mankind is left carrying an enormous intellectual burden that will live with us forever, namely Albert Einstein's ideas (as the only man to restructure physics), can never be properly understood by ordinary humans. They end in QED, whilst QED also shows the end of the study and creation of matter in physics. In addition, as the man who solved the problem of the origins

of time (and with time being what it is), Einstein was the world's only truly unique genius resembling God, set apart from humans, with ideas so advanced that they will always require interpretations by extraordinary scholars like Bertrand Russell, who was as clever as Aristotle without the latter's demerits. It is my sincere belief that however long human life survives on this planet, people will remember that I said this but was ignored even though it's true. For the avoidance of doubt, as the lawyers put it, my definition of time is this: in metaphysical jargons, time is the quantification of reality, existence or 'Being' reduced to shorter periodicities with mathematics and cyclical motions anywhere in the cosmos. Alternatively, in plain language, what we call time is part of reality 'out there', reduced to shorter periods with mathematics and cyclical motions to aid the management of existence so that we can live in the world in relative comfort and safety. This is the concept of time upon which the whole book is based.

References

With so many Footnotes, citing dozens of other books just to display learning is not my style. The books and papers cited below are the unavoidable ones for the rational discussion of time. I've used many of them in other books because my theory of time has not changed in fifty years!

ALBERT EINSTEIN (1879-1955) ---SPACE TIME, an article in the 1926/27 (13th) edition of the Encyclopedia Britannica. Also, RELATIVITY, in the same edition.

---NATURE No. 106, 782, (1921), almost the whole issue was devoted to the confirmation of Einstein's new theory of gravity.

---The Meaning of Relativity, Princeton University Press, 1966.

---The Evolution of Physics, (With Leopold Infeld) Cambridge 1838.

---RELATIVITY, Routledge Classics, London and New York, 2001.

HERMANN MINKOWSKI (1864-1909) ---He first mentioned his supposition in a lecture in cologne, known as Raum und Zeit (Space and Time) Cologne 21St September, 1908.

--- Herman Minkowski AdP 47, 927 (1915)

---Herman Minkowski, Goett. Nachr., 1908 p53. Reprinted in Gesammelte Abhandlungen von Herman Minkowski. Vol. 2, p352. Teubner, Leipzig 1911.

BERTRAND RUSSELL, FRS (1872-1970)---Our Knowledge of the External World, George Allen & Unwin, 1922.

--- Mysticism & Logic, George Allen & Unwin, 1976: a collection of important essays first published in 1917.

---ABC OF RELATIVITY, George Allen & Unwin, 1958 (recently revised by Professor Felix Pirani---first published in 1925.

---History of Western Philosophy, George Allen & Unwin, 1946.

---My Philosophical Development, George Allen & Unwin, 1958.

---The Analysis of Matter, George Allen & Unwin, 1927.

MORRIS KLINE: Mathematics in Western Culture, Allen & Unwin, London, 1954.

SIR ARTHUR STANLEY EDDINGTON, FRS (1862-1944)

---The Expanding universe, University of Michigan Press, Ann Arbor, 1933

---The Combination of Relativity Theory and Quantum Theory, Communication of the Dublin Institute of Advanced Studies, Dublin, 1943.

---The Mathematical Theory of Relativity, Cambridge, second ed. 1930.

---The Nature of the Physical World, Ann Arbor, Michigan, 1958.

---Philosophy of Physical Science, Cambridge, 1949.

---The Theory of Relativity and its Influence on Scientific Thought, Oxford, 1922.

---Space, Time and Gravitation, Cambridge, 1920.

SIR JAMES JEANS, FRS: Physics and Philosophy, Cambridge, 1942.

---The Mysterious Universe, Cambridge, 1930.

---The New Background of Science, Cambridge, 1933.

PROFESSOR A.N. WHITEHEAD: The Concept of Nature, Ann Arbor, Michigan, 1957.

---Science and the Modern World, Cambridge, 1922.

---An Inquiry Concerning the Principle of Natural Knowledge, Cambridge, 1919.

---Nature and Life, Cambridge, 1934.

---Process and Reality: An Essay in Cosmology, Cambridge, 1929.

---Essays in Science and Philosophy, Rider & Co., London, 1948.

---The Principle of Relativity, Cambridge, 1922.

Professor BANESH HOFFMANN: The strange Story of the Quantum, Dover Pub. Inc. New York, 1959.

Professor STEVEN F. SAVITT (ed.) Times Arrows Today: Recent Physical and Philosophical Work on the Direction of Time, Cambridge, 1995.

CHARLES A. FRITZ: Bertrand Russell's Construction of the External World, Routledge & Kegan Paul, London, 1952.

Professor JEREMY BERNSTEIN: Albert Einstein and the Frontiers of Physics, Oxford, 1996.

Professor RICHARD FEYNMAN: Lectures---The Character of Physical law. MIT Press, 1967. There are several volumes of the Feynman lectures and they are all worthy of serious study.

Abraham Pais, "Subtle is The Lord: The Life and Science of Albert Einstein", Oxford, 1982. Professor Pais has methodically provided details of almost all the original papers relevant to relativity. His list is so exhaustive I don't know of a better one anywhere.

WHAT REMAINS TO BE DISCOVERED, By Sir John Maddox, A Touchstone Book, Simon & Schuster, 1999.

Index

Abraham, 99, 108, 371
Alexander, 20
Archbishop, 105
Aristotle, xlv, 19, 26, 43, 100
astronomers, 83
astronomical, xxviii, 46, 77, 81, 88, 92, 104, 112, 113, 119
astronomy, xviii, xxxii, xxxiii, 4, 23, 25, 36, 41, 42, 54, 69, 73, 76, 84
astrophysics, xviii, xlii, xlv, 91
atomic, xviii, 27, 71, 87, 98, 102, 111
Bergson, xxx
Bernstein, 79

Bertrand, xxxv, xxxviii, xlii, xlviii, 13, 19, 26, 39, 43, 50, 52, 65, 66, 81, 82, 95, 100, 104, 371
Biblical, 33
Cambridge, xxxi, 19, 28, 39, 96, 99, 101, 368, 370, 371
Cartesian, xix
Copernicus, 98
Dualism, xix
Economics, 115
Eddington, xviii, xxx, xxxvi, xxxviii, xlii, xlviii, 8, 9, 21, 22, 26, 37, 40, 42, 45, 46, 47, 50, 51, 65, 67, 70, 75, 80, 99, 103, 104, 107

Einstein, xxvi, xxx, xxxi, xxxii, xxxv, xxxvi, xxxvii, xlii, xlv, xlvi, xlviii, xlix, lii, 7, 8, 10, 13, 14, 15, 18, 20, 21, 22, 26, 27, 28, 30, 37, 38, 39, 40, 41, 42, 43, 46, 50, 51, 54, 64, 67, 70, 74, 75, 76, 78, 79, 80, 82, 90, 92, 94, 98, 99, 100, 101, 102, 104, 105, 106, 107, 108, 109, 114, 368, 371
Einsteinian, xxxvii, 91
four-dimensional, 79, 105
Gottfried, 104
Hermann, 31
Leibniz, xxx, 35, 104
Lorentz, xxxv, xlii, xlviii, 12, 20, 26, 41, 64, 67, 75, 81, 82, 92, 95, 108, 113
Lorentz-Einstein, 41
Minkowski, xvii, xxxii, xxxiv, xxxvii, xxxviii, xl, 31, 54, 65, 79, 80, 83, 91, 100, 105, 107, 369
Moore, lii
Newton, xxix, 20, 44
Newtonian, 41
Nobel, xxxv, li, lii, 21, 43, 75

Palle, 104, 106
philosopher, xx, xlviii, 7, 18, 21, 26, 65, 66, 81, 100
Philosophy, 14, 83, 107, 369, 370, 371
Plato, xviii, li, lii, liii, 73, 94
Platonic, xxvi, 73
Professor, xvii, xviii, xxx, xxxvi, xxxviii, xlii, l, li, liii, 11, 28, 37, 42, 44, 45, 46, 47, 50, 65, 67, 69, 77, 79, 80, 92, 97, 99, 101, 103, 104, 105, 107, 116, 119, 369, 371
Pythagoras, 16, 105
Pythagorean, xxiv, 76
QED, xxvi, li, 38, 88, 92, 111
quantum, xix, xxvi, li, liii, 15, 21, 22, 38, 73, 74, 87, 92
Raum, 369
Rawlings, 44
relativity, xxx, xxxii, xxxiii, xl, xli, liii, 20, 22, 26, 28, 31, 32, 38, 50, 53, 76, 79, 80, 81, 82, 83, 86, 100, 101, 103, 106, 107, 371
Russell, xxx, xxxv, xxxviii, xlii, xliv, xlviii,

What Is Time?
xlix, l, li, lii, liii, 8, 10, 13, 14, 18, 24, 26, 27, 36, 38, 39, 40, 41, 42, 43, 47, 48, 50, 52, 65, 66, 67, 68, 70, 75, 80, 81, 82, 85, 92, 95, 98, 99, 100, 101, 104, 106, 108, 109, 111, 118, 120, 371

Russellian, xxxiv, 27
s=ct, xxxiii, xxxviii, 65
Stephenson, 20
Time Sequences, 109

University, xxxi, 39, 96, 368, 370
Whitehead, xvii, xxiv, xxx, xlii, xlix, l, li, liii, 8, 11, 20, 28, 32, 40, 42, 50, 52, 67, 68, 69, 70, 77, 89, 92, 97, 101, 104, 116, 119
Wittgenstein, xx, xxvii, 14, 18, 21, 66, 96
Yourgrau, 44, 104, 105, 106
Zeit, 369
Zero, 103

www.ingramcontent.com/pod-product-compliance
Lightning Source LLC
Chambersburg PA
CBHW060818170526
45158CB00001B/17